T0332169

A Beginner's Guide to MultiLevel Image Thresholding

Intelligent Signal Processing and Data Analysis
Series Editor: Nilanjan Dey

Intelligent signal processing (ISP) methods are progressively swapping the conventional analog signal processing techniques in several domains, such as speech analysis and processing, biomedical signal analysis radar and sonar signal processing, and processing, telecommunications, and geophysical signal processing. The main focus of this book series is to find out the new trends and techniques in the intelligent signal processing and data analysis leading to scientific breakthroughs in applied applications. Artificial fuzzy logic, deep learning, optimization algorithms, and neural networks are the main themes.

Bio-Inspired Algorithms in PID Controller Optimization
Jagatheesan Kallannan, Anand Baskaran, Nilanjan Dey, Amira S. Ashour

A Beginner's Guide to Image Preprocessing Techniques
Jyotismita Chaki, Nilanjan Dey

Digital Image Watermarking: Theoretical and Computational Advances
Surekha Borra, Rohit Thanki, Nilanjan Dey

A Beginner's Guide to Image Shape Feature Extraction Techniques
Jyotismita Chaki, Nilanjan Dey

Coefficient of Variation and Machine Learning Applications
K. Hima Bindu, Raghava Morusupalli, Nilanjan Dey, C. Raghavendra Rao

Data Analytics for Coronavirus Disease (COVID-19) Outbreak
Gitanjali Rahul Shinde, Asmita Balasaheb Kalamkar, Parikshit Narendra Mahalle, Nilanjan Dey

A Beginner's Guide to MultiLevel Image Thresholding
Venkatesan Rajinikanth, Nadaradjane Sri Madhava Raja, Nilanjan Dey

Hybrid Image Processing Methods for Medical Image Examination
Venkatesan Rajinikanth, E. Priya, Hong Lin, Fuhua (Oscar) Lin

For more information about this series, please visit: https://www.routledge.com/Intelligent-Signal-Processing-and-Data-Analysis/book-series/INSPDA

A Beginner's Guide to MultiLevel Image Thresholding

Authored by:

Dr. Venkatesan Rajinikanth
Dr. Nadaradjane Sri Madhava Raja
Dr. Nilanjan Dey

CRC Press
Taylor & Francis Group
Boca Raton London New York

CRC Press is an imprint of the
Taylor & Francis Group, an **informa** business

First edition published 2021
by CRC Press
6000 Broken Sound Parkway NW, Suite 300, Boca Raton, FL 33487-2742
and by CRC Press
2 Park Square, Milton Park, Abingdon, Oxon, OX14 4RN

ISBN: 978-0-367-50314-7 (hbk)
ISBN: 978-1-003-04944-9 (ebk)

Typeset in Times LT Std
by KnowledgeWorks Global Ltd.

Contents

Preface

Image processing widely adopted in a variety of domains to process various digital images obtained with a class of modalities. Improving the information existing in an unprocessed digital image is commonly performed with a chosen image enhancement technique. The image enhancement techniques play a major role in various domains, such as computer vision, pattern analysis, biometrics, data visualization, remote sensing, and medical image examination. Recently, image-assisted disease detection and treatment planning have improved healthcare. Based on this requirement, a number of semiautomated/automated disease evaluation tools are proposed and implemented. In medical image analysis, chosen preprocessing and post-processing techniques are implemented to extract and evaluate the disease infected section from the digital image. Further, the overall accuracy of this disease detection system depends on the preprocessing process. In this book, the commonly used image preprocessing technique called the multithresholding process is discussed with appropriate examples. The implementation of traditional and heuristic-algorithm-based thresholding techniques are also discussed with suitable examples. Further, a detailed study with thresholding techniques, such as Otsu, Kapur, Tsallis, Shannon, and Fuzzy-Tsallis are also presented. Finally, the implementation of the thresholding function is demonstrated using the lung CT scan slices of the COVID-19 dataset. In this book, the proposed work is experimentally demonstrated using MATLAB

The book is organized as follows:

Chapter 1 presents the overview of image enhancement techniques. The need for image enhancement is outlined, and well-known procedures, such as artifact removal, filtering, contrast enrichment, edge detection, thresholding, and smoothening are discussed briefly. The experimental demonstration of the proposed approach is presented using lung CT scan slice of COVID-19 database, and the simulation study is implemented using MATLAB.

Chapter 2 demonstrates the need for thresholding–bilevel and multilevel. Further, it presents the information about various thresholding techniques such as Otsu, Kapur, Tsallis, Shannon, and Fuzzy-Tsallis. Also, selection of an appropriate thresholding technique and its implementation requirements are discussed.

Chapter 3 presents the information regarding selection of appropriate image, differences between gray/RGB histograms, dimension complexity, and pixel distribution complexity. Image examination procedure is demonstrated using appropriate results attained using MATLAB software.

Chapter 4 shows the difference between the traditional thresholding and heuristic-algorithm-assisted thresholding procedures. Further, this chapter presents the details regarding Particle Swarm, Bacterial Foraging, Firefly, Bat, Cuckoo, and Social Group optimization algorithms and their role during the image thresholding process.

Chapter 5 presents the role of the objective function and the formation of the multiple objective functions to enhance the threshold outcome. Further, it presents the information on the image quality parameters to be computed to assess the performance of the executed threshold operation. An appropriate experiment is performed using the MATLAB, and the results are presented.

Chapter 6 demonstrates the processing of images with associated noise. In this chapter, grayscale/RGB images are stained with common noise functions, such as Gaussian, local variance, Poisson, salt & pepper, and speckle. The impact of noise on the image quality and threshold is discussed with appropriate experimental result.

Chapter 7 presents a detailed demonstration of the heuristic-algorithm-based image thresholding process using the well-known benchmark images. This work also presents essential guidelines for the selection of threshold function and objective function. Finally, the outcomes attained with these procedures are compared, and the suggestions to improve the demonstrated technique are presented.

Chapter 8 demonstrates the application of trilevel thresholding to improve 2D medical images recorded with various modalities. In this chapter, experimental investigation is implemented using the images recorded with MRI, CT, X-Ray, ultrasound, dermoscope, thermogram, and retinal fundus camera. The demonstrated results are obtained using MATLAB software.

Dr. Venkatesan Rajinikanth
St. Joseph's College of Engineering

Dr. Nadaradjane Sri Madhava Raja
St. Joseph's College of Engineering

Dr. Nilanjan Dey
JIS University, Kolkata, India

Acknowledgements

Dr. V. Rajinikanth and Dr. N. Sri Madhava Raja, editors of this book, would like to express their deep and sincere gratitude to **Dr. B. Babu Manoharan, M.A, M.B.A, Ph. D, Chairman, St. Joseph's College of Engineering,** (St. Joseph's Group of Institutions) OMR, Chennai. He has been their greatest exemplar ever. His vision, dynamism, and motivation have deeply inspired them always in their various accomplishments. Dr. Manoharan's encouragement for the teaching fraternity to excel in research have enthused them to explore various aspects of the same. The editors of this volume are extremely obliged to Dr. Manoharan for the help and motivation he extended in putting things together and successfully bringing out this book to share with the research community.

Author Biographies

Dr. Venkatesan Rajinikanth is a Professor in the Department of Electronics and Instrumentation Engineering at St. Joseph's College of Engineering, Chennai, India. Recently he edited a book titled *Advances in Artificial Intelligence Systems*, published by Nova Science Publishers, USA. He is the Associate Editor of *Int. J. of Rough Sets and Data Analysis* (IGI Global, US, DBLP, ACM dl) and edits special issues in the following journals: *Current Signal Transduction Therapy, Current Medical Imaging Reviews,* and *International Journal of Swarm Intelligence Research.* His main research interests include medical imaging, machine learning, and computer aided diagnosis as well as data mining.

Dr. Nadaradjane Sri Madhava Raja is passionate about teaching. He has 17 years of teaching experience at various engineering colleges. Currently, he is an Associate Professor at St. Joseph's College of Engineering, Chennai, India. He earned his doctorate in 2014, in the area of biomedical engineering. He completed his post-graduation in process control and instrumentation in 2002. His under graduate degree is in electrical and electronics engineering in 2001. Dr. Raja is also an ardent researcher and his major areas of research are medical image processing, optimization algorithms, heuristic algorithms, and biomechanics. He has published over 50 research papers in renowned journals and conference proceedings. He has also contributed chapters on optimization techniques to books published Nova Science Publishers, USA.

Dr. Nilanjan Dey is an Asso. Professor, Department of Computer Science and Engineering, JIS University, Kolkata, India. He is a visiting fellow of the University of Reading, UK. He was an honorary Visiting Scientist at Global Biomedical Technologies Inc., CA, USA (2012-2015). He was awarded his PhD. from Jadavpur Univeristy in 2015. He has authored/edited more than 70 books with Elsevier, Wiley, CRC Press and Springer, and published more than 300 papers. He is the Editor-in-Chief of International Journal of Ambient Computing and Intelligence, IGI Global, Associated Editor of IEEE Access and International Journal of Information Technology, Springer. He is the Series Co-Editor of Springer Tracts in Nature-Inspired Computing, Springer, Series Co-Editor of Advances in Ubiquitous Sensing Applications for Healthcare, Elsevier, Series Editor of Computational Intelligence in Engineering Problem Solving and Intelligent Signal processing and data analysis, CRC. His main research interests include Medical Imaging, Machine learning, Computer Aided Diagnosis, Data Mining etc. He is the Indian Ambassador of International Federation for Information Processing – Young ICT Group and Senior member of IEEE.

1 Introduction

1.1 INTRODUCTION TO IMAGE ENHANCEMENT

Normally, an image is used to provide the information with the help of picture distribution, and it presents the information in the form of a visual representation method. According to the registration, the image is classified as two-dimensional (2D) and multidimensional (3D), in which the processing procedures existing for the 2D images are quite simple compared to 3D. Furthermore, these images are further classified as conventional image (recorded with gray/RGB scale pixels) and binary image.

Ina variety of domains, images recorded using a chosen modality with a preferred pixel value can be used to deliver meaningful information. In some situations, the information existing in the unprocessed images is hard to understand and hence a number of pre-processing and post-processing procedures are proposed and implemented by the researchers [1, 2]. The implemented image processing schemes can help to improve the state of the raw image with a variety of methodologies, such as contrast-enhancement, edge-detection, noise-removal, filtering, fusion, thresholding, and segmentation [3].

Most of the existing enhancement procedures work well for gray-scale images compared to the RGB-scale images. In literature, a number of image examination procedures are available to pre-process test images and this process will help to convert a raw test image into an acceptable test image [4, 5]. The need for image enhancement and its practical significance is clearly discussed in the upcoming sections for both the grayscale and the RGB scale pictures.

1.2 IMPORTANCE OF IMAGE ENHANCEMENT

Normally, images recorded using a chosen image modality are referred to as unprocessed images. Based on the need, these raw images may be treated with chosen image conversion or enrichment techniques. Digital images recorded with well-known imaging methodologies may be associated with various problems. Before further assessment, it is essential to improve the information available in the image. In recent times, recorded digital images are processed and stored using digital electronic devices, and in order to ensure eminence, the image needs to be modified to convert the raw image into a processed image. Enhancement procedures such as (i) artifact removal, (ii) filtering, (iii) contrast enrichment, (iv) edge detection, (v) thresholding, and (vi) smoothening are some common procedures that are widely documented in literature are employed to convert the unprocessed image into a processed image. Image enhancement is essential to improve the visibility of the recorded information and to extract the recorded information from the enhanced image is quite easy compared to extracting it from the unprocessed image.

1.3   INTRODUCTION TO ENHANCEMENT TECHNIQUES

Currently, most of associated imaging systems are processor-controlled systems. Thereby, the images attained from these imaging systems are digital in nature. The quality of the image is judged based on the visibility of the region-of-interest (ROI) and the contrast between the background and ROI. An image recorded with a chosen imaging device is called a raw image. It is to be processed using a chosen image processing technique to convert the raw image into a usable image. This procedure is essential and a complex task when the image ROI has associated unwanted noise and artifact.

Hence, recently a number of image enhancement techniques are proposed and implemented by the researchers, and some of the commonly implemented methods are depicted below.

1.3.1 ARTIFACT REMOVAL

This procedure is essential to separate the image into a number of subsections based on a chosen threshold value. In order to group and extract the image pixels into various sections based on the chosen threshold value, the artifact removal technique usually implements a morphological filter along with a clustering approach. This approach is widely implemented in medical image processing applications to remove artifacts available in the 2D slices of magnetic resonance image (MRI) and computed tomography (CT). In order to demonstrate the chosen image enhancement technique (IET), this section considered the 2D CTI of COVID-19 patient obtained from the Radiopaedia database [6]. Figure 1.1 presents the structure of the procedure, and Figure 1.2 presents the results attained on the lung CTI of the COVID-19 image dataset (axial view).

Threshold filter is widely adopted to pre-process the medical images recorded with imaging modalities, such as MRI and CT. The earlier works on the MRI implemented the threshold-filter approach to separate the brain MRI slices into the skull section and normal brain anatomical regions. The lung CTI examination is also implemented with this technique to extract the lung and the abnormal image parts, such as heart and other body organs. After separating the test image with the chosen technique, the ROI is further considered for the examination process. The earlier works confirms that medical images (MRI as well as CT) pre-processed with this filter helped to attain better results compared to the unprocessed image [7, 8].

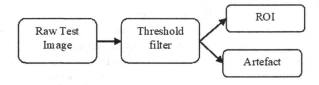

FIGURE 1.1   Implementation of threshold-filter based ROI extraction.

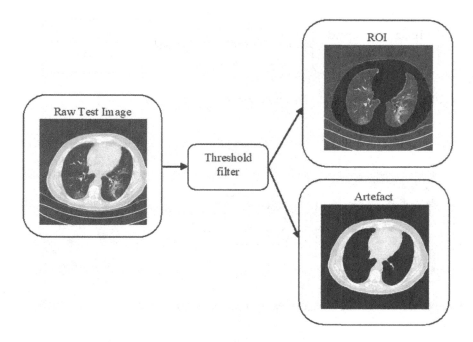

FIGURE 1.2   Experimental outcome with COVID-19 lung CT scan slice.

Advantages: Threshold filter helps to separate the ROI from the artifact and this process reduces the complexity in the test image to be evaluated.

Limitations: The main limitation of the threshold filter is the selection of finest threshold, which splits the test image into two sections. In most of the cases, the threshold selection is to be done manually with various trials. Trial based technique is a time-consuming process and this technique works only on the grayscale test image.

1.3.2   FILTERING

In the conventional digital signal processing domain, the filter implemented with a chosen technique and a preferred order is used to allow/block the signal information based on the frequency value. Similar to this operation, the unwanted pixels available in the digital image are removed/blocked using a chosen pixel filter.

Figure 1.3 presents the structure of the conventional filter, which eliminates the noise from the considered digital image of grayscale/RGB class. Figure 1.4 presents the results attained using the trial image of the COVID-19 database. In this work, the noise filter is employed to eliminate the noise (Salt and Pepper) associated with the test image. After the noise removal, a processed image with lesser pixel-level complexity is attained and it can be used for further enhancement using a chosen image processing technique to extract and evaluate the pneumonia infections seen in the image.

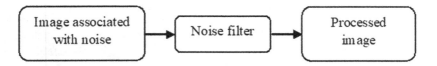

FIGURE 1.3   Execution of the filter to remove the associated noise.

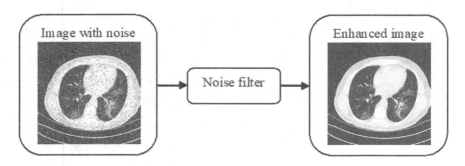

FIGURE 1.4   Removing the Salt & Pepper noise in the CT scan slice.

Advantages: The filter can be used to remove the unwanted pixels in the considered digital picture and it can be used as the pre-processing technique.

Limitations: Implementing a chosen image filter to enhance the test image is quite time-consuming, and the recent image processing procedure can process the test image even though it is corrupted with the noise.

1.3.3   CONTRAST ENRICHMENT

Normally, the information existing in the grayscale image is quite poor compared to the RGB scale image. In the grayscale test image, enhancing the ROI with respect to the background is a challenging task, and hence, a number of IET are proposed and implemented by the researchers to improve the ROI visibility. The contrast enhancement is one of the widely adopted techniques in medical image processing and this technique can be implemented in a number of ways. The methods such as histogram equalization, color-map tuning, and Contrast Limited Adaptive Histogram Equalization are some of the procedures considered to enhance the grayscale pictures [9].

Figures 1.5 and 1.6 present the structure of the IET and its related outcomes, respectively. From these figures, it can be observed that the results of the considered procedures will help to improve the visibility of the ROI compared to the raw image. The experimental investigation with the lung CTI confirms that the considered approaches improved the visibility of the COVID-19 infection considerably and later the improved ROI can be extracted and evaluated using a suitable segmentation procedure.

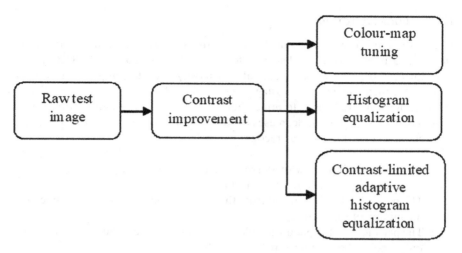

FIGURE 1.5   Commonly considered contrast enhancement methods.

Advantages: It is a straight forward approach and requires very less computational effort during the implementation.

Limitations: This approach is used during the initial level image enhancements, and in most of the image processing, it is an optional technique and the images associated with noise will not offer expected result with this technique.

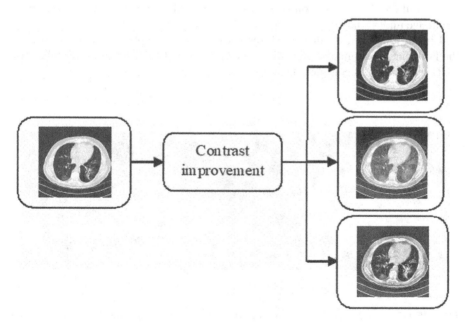

FIGURE 1.6   The experimental results attained with various contrast enhancement techniques.

1.3.4 EDGE DETECTION

This procedure is used to trace the borders of the ROI existing in the test image and literature; a number of edge detection techniques are available. The Canny Edge Detector (CED) is the most successful and mostly used technique from the year 1986 [10], which employs a multistage algorithm to identify a broad array of edges in the imagery. CED is used to mine valuable structural information from various images and, considerably, it decreases the amount of information to be investigated. The common condition for edge recognition includes following:

     a. Recognition of border with a small error rate, by accurately catching several edges existing in the considered test picture.
     b. The perimeter tip recognized from the operator should precisely confine on the middle of the border.
     c. The perimeter of the picture should only be noticed once, and the introduced image noise should not form fake boundaries.

To achieve the above said points, Canny implemented calculus of variations to optimize the operation. Sobel is also one of the common types of the edge detection procedure adopted in the image processing literature [11]. The edge detection result attained with Sobel and Canny on a lung CTI is presented in Figure 1.7. From Figure 1.7 (b), it can be noted that the experimental outcome of the Canny is superior to Sobel and the choice of particular edge detection can be chosen as per the expertise of the operator.

Advantage: Edge detection is essential to detect the boundary and the texture of the image under study.

Limitation: The edge detection procedure requires complex procedures to detect the boundary of an image, and this procedure will not offer fitting result when the RGB scale image is considered.

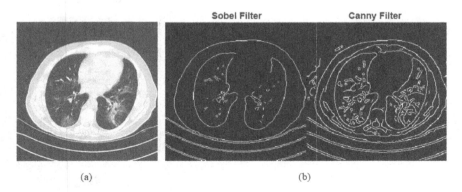

FIGURE 1.7 Result attained with the edge detection technique. (a) Test image, (b) attained result with Sobel and Canny filters.

1.3.5 THRESHOLDING

Image ROI improvement based on the threshold operation is widely adopted in the literature to process a class of traditional and medical images. In most of the image examination systems, thresholding is adopted as the pre-processing approach. In the thresholding process, the ROI in a chosen image (grayscale/RGB) is enhanced by grouping the pixels based on the selected finest threshold value. The grouping of the image pixel will separate the ROI from the image background and other parts, and after this enhancement, the ROI can be extracted using a chosen segmentation process.

In the literature, the thresholding can be achieved using the traditional procedures and soft-computing driven techniques. In the traditional/operator-assisted methods, the finest threshold identification is manually identified using various trials. During this, the threshold of the test image is rapidly varied using its histogram, and during this operation, the different pixel grouping is studied to attain the enhanced image as per the requirement. Identification of the finest threshold using the conventional procedure is complex and time-consuming. To overcome this problem, recently, the threshold operation is performed under the supervision of soft-computing algorithms [12]. In hybrid medical data assessment examination techniques, thresholding is adopted as the pre-processing method to enhance the ROI [13].

Based on the chosen number of thresholds, it is classified as bilevel (separating the image into ROI and background) and multilevel threshold (dividing them into various pixel groups). The threshold result obtained on a chosen lung CTI of the COVID-19 database is depicted in Figure 1.8 (a) and (b) and this result verifies that the threshold process will improve the visibility of the pneumonia infection in the lung CTI.

Advantages: Thresholding is a simple and effective IET and it can be implemented manually and with the help of heuristic algorithms. Further, it supports the processing of the grayscale/RGB scale images using varied schemes. ROI of the test image can be improved with either a bilevel or a multilevel threshold selection process.

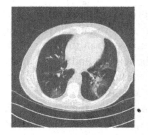  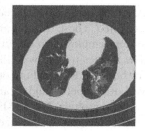

(a) Test image   (b) Bi-level threshold   (c) Multi-level threshold

FIGURE 1.8 Multilevel threshold implemented on the lung CT scan slice.

Limitations: The choice of the finest threshold and the selection of the objective function during the soft-computing assisted thresholding are quite difficult. Further, in most of the image enhancement techniques, the thresholding is chosen only as the pre-processing technique and the computation time of this operation increases based on the histogram complexity and the pixel distribution.

1.3.6 SMOOTHENING

During automated detection and classification operations, texture and shape features extracted from the image plays a vital role. Before extracting texture features from an image, it is necessary to treat the raw image with preferred image normalization procedures, and the image's texture smoothening based on a chosen filter is widely adopted to improve the texture features. The earlier work confirmed that the Gaussian-Filter (GF) technique can be employed to improve the texture and edge features of the grayscale image of a chosen dimension [14]. Further, one of the earlier research works confirmed that the GF with varied scale (ϕ) will provide the enlargement of texture pattern in vertical and horizontal directions [15].

The conventional Gaussian operator for a 2D image case is presented as in Eqn. 1.1

$$U(x,y) = \frac{1}{2\pi\phi^2} e^{-\left(\frac{x^2+y^2}{2\phi^2}\right)} \tag{1.1}$$

where ϕ = standard deviation and $U(x,y)$ = Cartesian coordinates of the image. By altering ϕ, we can create imagery with diverse edge enhancements.

Laplacian of a Gaussian function as a filter, which helps to notice edges by resulting the zero-crossings of their second derivatives as in Eqn.1.2

$$\nabla^2 U(x,y) = \frac{d^2}{dx^2} U(x,y) + \frac{d^2}{dy^2} U(x,y) = \frac{x^2+y^2-2\phi^2}{2\pi\phi^6} e^{-\left(\frac{x^2+y^2}{2\phi^2}\right)}. \tag{1.2}$$

Figure 1.9 presents the chosen lung CTI and the related experimental results. Figure 1.9 (b) presents the horizontally smoothened lung CTI for a chosen value of $\phi = 60$ and Figure 1.9 (c) depicts the vertically smoothened image for $\phi = 300$. After the possible enhancement, the texture and edge features from these images are extracted. These features are then considered to train, test, and validate the classifier system employed to detect/classify the COVID-19 infection seen in the lung CTI. The images treated with the GF will show a variety of the texture and edge values for both the ROI and the other sections of the image and evaluation of these values will offer a better understanding of the images and its abnormalities.

Advantages: The GF-based techniques are used to normalize the texture and edge of the grayscale images of varied dimensions. Further, this method can be used to generate various edges as well as texture patterns based on the chosen ϕ. Further, the GF supports a variety of the edge and texture detection operations. The Canny edge detection technique also employs the GF to improve the edges of the test image.

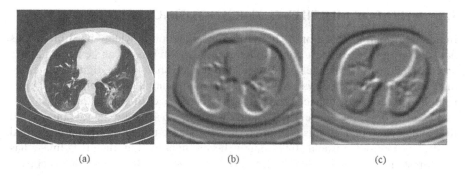

(a)   (b)   (c)

FIGURE 1.9   Results attained with Gaussian-filter based enhancement. (a) Trial image, (b) image with horizontal smoothening, (c) image with vertical smoothening.

Limitations: The information existing in the GF-filter-treated test image can be examined only with a chosen image examination technique and the information cannot be examined with manual operators. Hence, this IET can be used only when a computerized algorithm is employed to detect/classify the image based on its ROI's edge/texture value.

1.4   RECENT ADVANCEMENTS

In earlier days, the image examination is performed by an experienced operator, and the attained outcome by this technique requires some compromise. In the current era, due to the rapid advancement in the modern technology, usage of computers becomes very common in almost all domains due to its simplicity and adaptability. Recently a considerable number of computer algorithms are developed to support a variety of image examination operations. Further, the computerized image processing supports: (i) semi-automated/automated examination, (ii) works well for grayscale/RGB images, (iii) can be used to implement a variety of soft-computing techniques, and (iv) results of these methods can be stored temporarily or permanently based on the requirement.

The computerized algorithms are helped to implement a variety of image processing procedures to enhance the quality and the information of the images irrespective of their color, size, and pixel distribution. Further, recently number of software are developed to support the computerized image processing procedures without compromising their quality, and the throughput attained with the computerized methods are more compared to the operator assisted methods.

Due to its merits, recently computer-assisted procedures are implemented to perform the essential operations, such as artifact removal, contrast enhancement, edge detection, thresholding, and smoothening. Further, the availability of the heuristic algorithms and their support toward the image processing has also helped to develop and implement different machine-learning and deep-learning technique to process a variety of images with a preferred pixel dimension.

1.5 NEED FOR MULTILEVEL THRESHOLDING

Image thresholding based on a chosen guiding function is widely used in various fields to pre-process the test image to be examined. An image can be known as the arrangement of various pixels with respect to the thresholds. In a digital image, the distribution of the pixel plays a major role, and modification or grouping of the pixel is highly preferred to enhance/change the information existing in the image.

In earlier days, the bilevel thresholding is chosen to separate the raw image into ROI and background. In this operation, the operator/computer algorithm is allowed to identify Finest Threshold (FT) using a chosen function. Let, the given image has 256 thresholds (ranging from 0 to 255) and from this, one threshold value is chosen as the finest threshold as follows:

$$0 < \text{Finest threshold} < 255. \tag{1.3}$$

The chosen threshold will help to separate the image into two sections: section 1 have the pixel distribution < FT and section 2 with pixel distribution > FT. This operation separates the image into two-pixel groups and, hence, it is called as the bilevel thresholding.

If the aim is to separate the given image into more than two image sections, then a multilevel thresholding is preferred. During this operation, the number of FT = number of threshold levels. For example, if a trilevel threshold is chosen for the study, it will separate the test image into three sections as depicted in Eqn 1.4

$$0 < FT_1 < FT_2 < 255. \tag{1.4}$$

In this case, the image is separated into three sections: section1 (pixels between 0 and FT_1), section 2 (pixels between FT_1 and FT_2), and section3 (pixels between FT_2 to 255). In most of the applications, the information attained with the bilevel threshold is not appropriate, and hence, a multithresholding is referred to pre-process grayscale/RGB images. The information on multilevel threshold can be found in the literature [16, 17].

1.6 IMPLEMENTATION AND EVALUATION OF THRESHOLDING PROCESS

Optimal-threshold-based IET can be executed using a manual operator or by using a computer algorithm. The computation complexity in operator-assisted FT selection process needs more computation effort. Further, this complexity will increase if the number of FT to be identifier is >2 (i.e., multithreshold selection). Since, the operator is to adjust the thresholds arbitrarily till the best image separation is achieved. Due to this reason, in recent years, heuristic-algorithm-assisted image thresholding is widely proposed by the researchers. In this method, a chosen heuristic algorithm with a chosen objective function is implemented to enhance the image information.

After implementing the IET using a chosen procedure, its outcome is to be validated in order to verify the eminence of the proposed procedure. The validation of

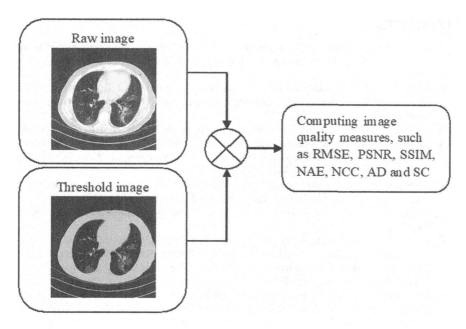

FIGURE 1.10   Performance validation of the thresholding process.

the thresholding performance can be achieved with a comparative analysis between the raw image and threshold image. During this process, commonly considered image quality values, such as Root Mean Squared Error, Peak Signal-to-Noise Ratio (in dB), Structural Similarity Index, Normalized Absolute Error, Normalized Cross Correlation, Average Difference, and Structural Content, are computed for every image, and based on this value, the performance of the implemented thresholding technique is confirmed. Further, the results attained with a chosen technique are to be compared against the results of state-of-the-art techniques existing in the literature [18–21]. The commonly employed evaluation procedure to verify the result of the threshold process is depicted in Figure 1.10 and based on its outcome, the eminence of the implemented procedure is confirmed.

1.7   SUMMARY

The need for image processing systems and implementation of various possible image enhancement procedures to modify the raw test image is outlined with appropriate results. Further, the need for image thresholding and the implementation of bilevel and multilevel thresholding is clearly discussed. Implementation of thresholding process for the grayscale/RGB images and its evaluation procedures are presented. From the above discussion, it can be noted that image thresholding performed using a chosen technique can be implemented to pre-process the digital test images recorded with a class of image modalities. Moreover, the image thresholding has emerged as a key research field due to its practical significance.

REFERENCES

1. Raja, N.S.M., Rajinikanth, V., & Latha, K. (2014). Otsu based optimal multilevel image thresholding using firefly algorithm. *Modelling and Simulation in Engineering, 2014,* 794574.
2. Akay, B. (2013). A study on particle swarm optimization and artificial bee colony algorithms for multilevel thresholding. *Applied Soft Computing Journal, 13*(6), 3066–3091.
3. Dougherty, (1994)). *Digital Image Processing Methods,* 1st Edition, CRC Press, USA.
4. Satapathy, S.C., Raja, N.S.M., Rajinikanth, V., Ashour, A.S., & Dey, N. (2018). Multilevel image thresholding using Otsu and chaotic bat algorithm. *Neural Computing and Applications, 29*(12),1285–1307.
5. Rajinikanth, V., & Couceiro, M.S. (2015). RGB histogram based color image segmentation using firefly algorithm. *Procedia Computer Science, 46,* 1449–1457.
6. Available online: https://radiopaedia.org/cases/covid-19-pneumonia-27 (Accessed on 16th August 2020).
7. Bhandary, A., Prabhu, G.A., Rajinikanth, V., Thanaraj, K.P., Satapathy, S.C., Robbins, D.E., Shasky, C., Zhang, Y.D., Tavares, J.M.R.S., & Raja, N.S.M. (2020). Deep-learning framework to detect lung abnormality–A study with chest X-Ray and lung CT scan images. *Pattern Recognition Letters, 129,* 271–278.
8. Rajinikanth, V., Dey, N., Raj, A.N.J., Hassanien, A.E., Santosh, K.C., & Raja, N.S.M. (2020). Harmony-search and otsu based system for coronavirus disease (COVID-19) detection using lung CT scan images. arXiv:2004.03431.
9. Stark, J.A. (2000). Adaptive image contrast enhancement using generalizations of histogram equalization. *IEEE Transactions on Image Processing, 9*(5),889–896.
10. Canny, J. (1986). A computational approach to edge detection. *IEEE Transactions on Pattern Analysis and Machine Intelligence,* 8(6), 679–698.
11. Zhou, P., Ye, W., & Wang, Q. (2011). An improved canny algorithm for edge detection. *Journal of Computational Information Systems, 7*(5), 1516–1523.
12. Dey, N., et al (2019). Social-group-optimization based tumor evaluation tool for clinical brain MRI of Flair/diffusion-weighted modality. *Biocybernetics and Biomedical Engineering, 39*(3),843–856.
13. Fernandes, S.L., Rajinikanth, V., & Kadry, S. (2019). A hybrid framework to evaluate breast abnormality using infrared thermal images. *IEEE Consumer Electronics Magazine, 8*(5), 31–36.
14. Basu, M. (2002). Gaussian-based edge-detection methods-a survey. *IEEE Transactions on Systems, Man, and Cybernetics, Part C: Applications and Reviews, 32*(3), 252–260. Available online:https://doi.org/10.1109/TSMCC.2002.804448.
15. Marr, D. & Hildreth, E.. (1980). Theory of edge detection. *Proceeding of the Royal Society of Londondon Series A, Mathematical and Physical Sciences, 207,* 187–217.
16. Ghamisi, P., Couceiro, M.S., Benediktsson, J.A., & Ferreira, N.M.F. (2012). An efficient method for segmentation of images based on fractional calculus and natural selection. *Expert System with Appliaction, 39*(16), 12407–12417.
17. Ghamisi, P., Couceiro, M.S., Martins, F.M.L., & Benediktsson, J.A. (2014). Multilevel image segmentation based on fractional-order Darwinian particle swarm optimization. *IEEE Transaction on Geoscience and Remote sensing,52*(5), 2382–2394.
18. Manikantan, K., Arun, B.V., & Yaradonic, D.K.S. (2012). Optimal multilevel thresholds based on tsallis entropy method using golden ratio particles warm optimization for improved image segmentation. *Procedia Engineering, 30,* 364–371.

19. Rajinikanth, V., Raja, N.S.M., & Latha, K. (2014). Optimal multilevel image thresholding: An analysis with PSO and BFO algorithms. *Australian Journal of Basic and Applied Sciences*, *8*(9), 443–454.
20. Hore, A. & Ziou, D. (2010). Image Quality Metrics: PSNR vs. SSIM. In: *IEEE International Conference on Pattern Recognition (ICPR)*, 2366–2369. Turkey, Istanbul.
21. Wang, Z., Bovik, A.C., Sheikh, H.R., & Simoncelli, E.P. (2004). Image quality assessment: From error measurement to structural similarity. *IEEE Transactions on Image Processing*, *13*(1), 1–14.

2 Thresholding Approaches

2.1 NEED FOR IMAGE THRESHOLDING

Thresholding is a proven image preprocessing procedure widely adopted in the image processing literature to improve the visibility of regions of interest (ROI) of the test images. For the chosen test image (grayscale/RGB), the threshold value can be obtained by plotting histogram of the image. The histogram is a graphical representation of the pixel distribution (X-axis) of the image with respect to the threshold value (Y-axis). For simplicity, the normally considered threshold value during the image examination task is chosen as $L = 256$. For every image, the threshold value is fixed and pixel distribution will vary based on the size and information available in the digital image [1, 2].

Let us consider the lung CT scan images presented in Figure 1.2(a) for discussion. The dimension of the image is $256 \times 256 \times 1$ pixels, and the histogram of the image represents distribution of the image pixels with respect to the threshold [3, 4]. For this image, enhancement of the COVID-19-infected region is achieved by employing bilevel and multilevel threshold selection processes and results are clearly presented in Figure 1.2(b). Other related information regarding the thresholding can be found in following research articles [5–8].

The image threshold process helps to improve the pixel-level information available in digital information, which can be recognized and analyzed for further assessment. In the medical domain, thresholding process will separate the digital image into background, normal section, and section with disease. Finally, the section with disease-related information can be extracted with a chosen technique and examined further. As mentioned earlier, thresholding is one of the well-known preprocessing techniques adopted in the medical image evaluation domain [9–11].

2.2 BILEVEL AND MULTILEVEL THRESHOLD

The aim of the image thresholding is to identify the finest/optimal threshold, which separates the image into various classes based on the assigned threshold requirement. The threshold selection for the grayscale images is straightforward compared to RGB, since in gray scale case, implementation is easy but for RGB case, it is separately implemented for histograms of R, G, and B, respectively. According to the number of thresholds, it is classified as (i) bilevel and (ii) multilevel threshold, and this procedure is clearly discussed further in this chapter with appropriate experimental results attained with the MATLAB software.

For simplicity, we will discuss the thresholding of bilevel case and it can be extended to the multilevel case with appropriate modifications. Let, $T = (t_0, t_1, ..., t_{L-1})$ represents the number of thresholds available in a chosen digital image of fixed size. Every image is to be assessed by considering its histogram formed by considering its pixel distribution (Y-axis) and threshold distribution (X-axis). The thresholding process needs the identification of an optimal threshold value $T = t_{OP}$, which supports the grouping of image pixels in order to enhance the visibility of the ROI. The thresholding scheme can be implemented using (i) bilevel approach (sorting out image pixels into two groups) and (ii) multilevel technique (sorting out image pixels into multiple clusters).

- Bilevel threshold: This procedure helps to separate the test image into two pixel groups, i.e., ROI and background based on $T = t_{OP}$. Level 1 will be obtained by considering image pixels $>T = t_{OP}$ and level 2 will be obtained by considering image pixels $<T = t_{OP}$, and to achieve this, a manual operator is employed who can identify the $T = t_{OP}$ using a trial and error approach. On the other hand, a heuristic algorithm can be used to solve this problem with lesser effort.
- Multilevel threshold: This is an extension of the bilevel operation, in which the considered digital photography is separated into more than two pixel groups. During this operation, a considerable number of t_{OP} is identified. If an image is separated with $T = t_{OP1}, ... t_{OPn}$, then it is termed as the multi-thresholding process.

2.3 COMMON THRESHOLDING METHODS

Identification of a suitable technique to enhance the digital image is a quite challenging task. An appropriate image examination technique can be chosen based on the recommendation by earlier works or experience. In the literature, a number of threshold techniques are available, and commonly considered approaches are represented in Figure 2.1. In the literature, the histogram-assisted threshold selection procedure is widely considered to enhance the grayscale/RGB scale picture compared to their alternatives. Each method has its own merits and demerits.

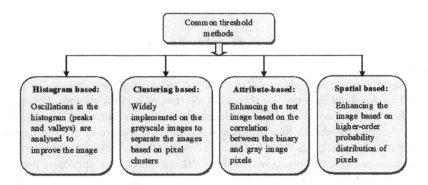

FIGURE 2.1 Common threshold selection procedures for image enhancement.

In the histogram-assisted threshold process, the image histogram is examined using a chosen technique, in which threshold values are randomly varied till the quality of the processed image reaches to a certain level. In this technique, a chosen objective value is taken as the guiding mechanism to justify the quality of image based on a chosen image-related measure. This measure is called the objective function.

In the histogram-based technique, the complexity of the thresholding increases due to (i) size of the image, (ii) pixel distribution, and (iii) multiple pixel class (i.e., RGB). When image dimensions increase, number of pixels in the image will also rise. As a result, this may increase the computation time during threshold selection. Furthermore, thresholding complexity will also increase with an uneven pixel distribution (large peaks and valleys in histogram) and the RGB class histogram as well.

Usually, every thresholding operation needs a monitoring parameter to guide the threshold selection process and maximization of the monitoring parameter. This is always considered in the literature to enhance the quality of test images [7, 8]. Even though a considerable number of monitoring techniques exist in the literature, few techniques are largely implemented in the image thresholding operation due to their superiority. According to these functions, these techniques can be classified as (i) between-class-based techniques and (ii) entropy-based methods.

Between-class-assisted image thresholding was proposed by Otsu, and due to its eminence, a number of research works have employed the maximization of Otsu's function to identify optimal threshold during bilevel and multilevel operations. Further, the maximization of entropy-based methods is also widely adopted in the literature to process a class of images. Upcoming sections present details on commonly used objective functions in the literature to threshold the chosen image.

2.3.1 OTSU'S APPROACH

Otsu's technique was primarily discussed in 1979. It works on the basis of between-class-variance concept and identifies the finest threshold by maximizing the objective value [12]. The between-class-variance is Otsu's nonparametric threshold selection concept, to be computed by randomly varying the image threshold using a chosen technique. During this process, the selection of $T = t_{OP}$ is essential to convert the raw image into processed image. This procedure works well on bilevel and multilevel threshold processes and its implementation is depicted in Figure 2.2.

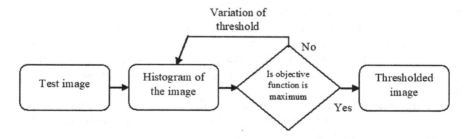

FIGURE 2.2   Optimal threshold selection using Otsu's function.

Let us consider the gray level image case for discussion. During the bilevel operation, $T = t_o, t_1$ are selected, which divides the input image into two groups, i.e., G_0 and G_1 (image background and ROI). The group G_0 contains gray pixels of range 0 to t_0 and class G_1 encloses gray levels from t_1 to 225 [13].

This function can be mathematically expressed on the basis of its probability distribution function, and its distribution for gray levels G_0 and G_1 can be denoted as

$$G_0 = \frac{P_0}{\eta_0(T)} \cdots \frac{P_{t_0-1}}{\eta_0(T)} \quad \text{and} \quad G_1 = \frac{P_{t_0}}{\eta_1(T)} \cdots \frac{p_{255}}{\eta_1(T)} \tag{2.1}$$

where $\eta_0(T) = \sum_{i=0}^{T-1} P_i$, $\eta_1(T) = \sum_{i=T}^{255} P_i$.

The mean values ψ_0 and ψ_1 for G_0 and G_1 can be denoted as:

$$\psi_0 = \sum_{i=0}^{T-1} \frac{iP_i}{\eta_0(T)} \quad \text{and} \quad \psi_1 = \sum_{i=T}^{255} \frac{iP_i}{\eta_1(T)} \tag{2.2}$$

The mean intensity (ψ_T) of the entire image can be represented as:

$$\psi_T = \eta_0 \psi_0 + \eta_1 \psi_1 \quad \text{and} \quad \eta_0 + \eta_1 = 1.$$

The objective function for bilevel thresholding problem can be expressed as:

$$\text{Otsu}_{\max} = J(T) = \vartheta_0 + \vartheta_1 \tag{2.3}$$

where $\vartheta_0 = \eta_0 (\psi_0 - \psi_T)^2$ & $\vartheta_1 = \eta_1 (\psi_1 - \psi_T)^2$.

This technique can be modified to a multilevel threshold problem by including various "T" values as follows.

Let us consider a chosen test image has thresholds distributed as: $T = (t_0, t_1, ..., t_{L-1})$, which helps to divide the chosen image into multiple thresholds such as G_0 with gray thresholds 0 to t_0, G_1 with gray thresholds t_0 to t_1, . . ., G_T with gray thresholds t_T to 255

The objective function for multilevel thresholding can be defined as

$$\text{Otsu}_{\max} = J(T) = \vartheta_0 + \vartheta_1 + \cdots + \vartheta_{L-1} \tag{2.4}$$

where $\vartheta_0 = \eta_0 (\psi_0 - \psi_T)^2$, $\vartheta_1 = \eta_1 (\psi_1 - \psi_T)^2$, ..., $\vartheta_T = \eta_T (\psi_T - \psi_{L-1})^2$.

Based on the requirement, the threshold value can be chosen as $T = 2, 3, 4, ... L - 1$.

Advantages: Otsu's technique is one of the common and frequently used techniques to process a variety of images. This approach works well on grayscale images and helps to attain better values of image quality parameters compared to other related techniques.

Limitations: Even though this technique works on a variety of images, its performance is less as compared to the entropy-based technique when abnormality of the image is the prime ROI. When abnormality is to be examined, the entropy-assisted technique offers better results compared to Otsu's technique.

2.3.2 Tsallis Approach

In general, entropy is connected with the calculation of disorder in an arrangement. Shannon originally measured the entropy-based assessment to calculate the uncertainty concerning in sequence substance of the scheme [14]. Shannon assured that when a substantial structure is detached as two statistically free subsystems F_1 and F_2, then its entropy can be expressed as [15]

$$S_h(F_1 + F_2) = S_h(F_1) + S_h(F_2). \tag{2.5}$$

From the above equation, a non-extensive entropy-based concept was introduced by Tsallis as shown below:

$$S_{hQ} = \frac{1 - \sum_{i=1}^{T}(P_i)^Q}{Q - 1} \tag{2.6}$$

where T = threshold and Q = entropic index.

Eq. (2.5) will meet Shannon's entropy (SE) when $Q \to 1$.

The entropy value can be expressed with a pseudo additive rule as:

$$S_{hQ}(F_1 + F_2) = S_{hQ}(F_1) + S_{hQ}(F_2) + (1 - Q) \cdot S_{hQ}(F_1) \cdot S_{hQ}(F_2). \tag{2.7}$$

Tsallis' technique is then considered to find T of the chosen image. Let the test image has L gray values of span $\{0, 1, 2, ..., L\}$ with a probability functions, $P_i = P_0, P_1, ..., P_{L-1}$, for $L = 256$.

For the multilevel case, it can be presented as:

$$\text{Tsallis}(t_i) = [t_0, t_1, ..., t_{L-1}] = \arg\max[S_{hQ}^{F_1}(t) + S_Q^{F_2}(t) + \cdots + S_Q^{k}(t)$$
$$+ (1 - Q) \cdot S_Q^{F_1}(t) \cdot S_Q^{F_2}(t), ..., S_Q^{F_K}(t)] \tag{2.8}$$

where

$$S_{hQ}^{F_1}(t) = \frac{1 - \sum_{i=0}^{t_1-1}\left(\frac{P_i}{P^{F_1}}\right)^Q}{Q - 1}, \quad P^{F_1} = \sum_{i=0}^{t_1-1} P_i$$

$$S_{hQ}^{F_2}(t) = \frac{1 - \sum_{i=t_1}^{t_2-1}\left(\frac{P_i}{P^{F_2}}\right)^Q}{Q - 1}, \quad P^{F_2} = \sum_{i=t_1}^{t_2-1} P_i \tag{2.9}$$

$$S_Q^{F_k}(t) = \frac{1 - \sum_{i=t_k}^{L-1}\left(\frac{P_i}{P^k}\right)^Q}{Q - 1}, \quad P^{F_K} = \sum_{i=t_k}^{L_2-1} P_i.$$

Subject to the following constraints:

$$\left| P^{F_1} + P^{F_2} \right| - 1 < S_h < 1 - \left| P^{F_1} - P^{F_2} \right|$$

$$\left| P^{F_2} + P^{F_3} \right| - 1 < S_h < 1 - \left| P^{F_2} - P^{F_3} \right| \tag{2.10}$$

$$\left| P^{F_k} + P^{F_{L-1}} \right| - 1 < S_h < 1 - \left| P^{F_k} - P^{F_{L-1}} \right|.$$

During multilevel thresholding, the aim is to discover the best threshold value that maximizes Tsallis(t_i).

Advantages: Entropy-assisted technique works well on a class of images cases and, particularly, it offers better results during the image abnormality examination task.

Limitations: Compared with other approaches, the image quality measures obtained with the Tsallis function are low. Further, the grouping of pixel achieved with this technique is poor compared to Shannon's technique.

2.3.3 FUZZY-TSALLIS ENTROPY APPROACH

Fuzzy-Tsallis (FT) entropy is derived from the Tsallis function and is considered to solve bilevel and multilevel thresholding problems [16, 17]. The earlier work confirms that this approach offers enhanced image quality compared with the Tsallis technique.

The mathematical description of FT entropy is expressed as described in below given text.

Let us consider an image (*I*) of dimension $A \times B$ with the maximum threshold value of $L = 256$.

Then, $I = \{(i, j) : i = 0, 1, ..., A - 1; j = 0, 1, ..., B - 1\}$ for thresholds $t = 0, 1, ..., L - 1$.

where A, B, and L are +ve numbers.

Let $I_G(X, Y)$ be the gray level value of test image at pixel (X, Y), then

$$I_t = \{(X, Y) : I_G(X, Y) = t, (X, Y) \in I\}. \tag{2.11}$$

Let us consider bilevel thresholding for the discussion which separates the image into two sections: $I = I_1, I_2$.

For an image (I), the probability distribution can be expressed as:

$$\prod_2 = \{I_1, I_2\} \tag{2.12}$$

$$p_3 = P(I_3), \; p_1 = P(I_1) \tag{2.13}$$

where I_1 = darker pixel group (denoted as D or 1), I_2 = brighter pixel group (denoted as B or 2), and p depicts the probability distribution.

For every t, the distribution can be stated as:

$$\left. \begin{aligned} I_{t1} &= \{(X,Y): I_G(X,Y) \le t, (X,Y) \in I_t\} \\ I_{t2} &= \{(X,Y): I_G(X,Y) > t, (X,Y) \in It\} \end{aligned} \right\} \tag{2.14}$$

$$\left. \begin{aligned} p_{t1} &= P(I_{t1}) = p_t * p_{1/t} \\ p_{t2} &= P(I_{t2}) = p_t * p_{2/t} \end{aligned} \right\}. \tag{2.15}$$

In Eq. (2.15), $p_{1/t}$ and $p_{2/t}$ are conditional probabilities of a pixel which are grouped as dark and bright, with respect to I_t with $p_{1/t} + p_{2/t} = 1$ $(t = 0,1,...,255)$.

Let $\eta_1(t)$ and $\eta_2(t)$ are grades for dark and bright pixel groups, then

$$\left. \begin{aligned} p_1 &= \sum_{t=0}^{255} p_t * p_{1/t} = \sum_{t=0}^{255} p_t * \eta_1(t) \\ p_2 &= \sum_{t=0}^{255} p_t * p_{2/t} = \sum_{t=0}^{255} p_t * \eta_2(t) \end{aligned} \right\}. \tag{2.16}$$

The grade $\eta_1(t)$ and $\eta_2(t)$ can be attained using fuzzy membership functions $M_{F1}(t,a,c)$ and $M_{F2}(t,a,b,c)$ correspondingly. It is presented in Figure 2.3, and its expression is presented as:

$$\eta_1(t) = \begin{cases} 1, & t \le a \\ 1 - \dfrac{(t-a)^2}{(c-a)*(b-a)}, & a < t \le b \\ \dfrac{(t-c)^2}{(c-a)*(c-b)}, & b < t \le c \\ 0, & t > c \end{cases} \tag{2.17}$$

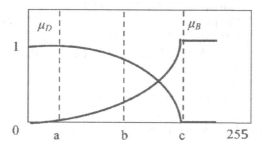

FIGURE 2.3 Fuzzy membership function for a chosen image.

$$\eta_2(t) = \begin{cases} 0, \\ \dfrac{(t-a)^2}{(c-a)*(b-a)}, & t \le a \\ & a < t \le b \\ 1 - \dfrac{(t-c)^2}{(c-a)*(c-b)}, & b < t \le c \\ & t > c \\ 1, \end{cases} \tag{2.18}$$

where the position of a, b, and c will be between thresholds; $0 \le a \le b \le c \le 255$.
For every pixel cluster, the FT can be expressed as

$$S_1 = \dfrac{1 - \displaystyle\sum_{t=o}^{255}\left(\dfrac{p_k * \eta_1(k)}{p_1}\right)^Q}{Q-1} \\ S_2 = \dfrac{1 - \displaystyle\sum_{t=o}^{255}\left(\dfrac{p_k * \eta_2(t)}{p_2}\right)^Q}{Q-1} \Bigg\}. \tag{2.19}$$

Finally, total FT can be represented as conventional Tsallis function:

$$S_Q(1+2) = S_Q(1) + S_Q(2) + (1-Q)*S_Q(1)*S_Q(2) \tag{2.20}$$

$$\text{Tsallis}_{max}(T) = \arg\max[S_Q^1(T) + S_Q^2(T)(1-Q).S_Q^1(T).S_Q^2(T)]. \tag{2.21}$$

Advantages: This procedure works well during the medical image evaluation task and separates the disease section from image background very distinctly.

   Limitations: Implementation of this technique is quite complex compared to the traditional Tsallis technique.

2.3.4   SHANNON'S APPROACH

Description of SE was discussed by Kannappan [18]. In the SE, let us choose a test picture of size $X \times Y$. The pixel association in test picture (x, y) is defined as $F(x, y)$, for $x \in \{1,2,...,X\}$ & $y \in \{1,2,...,Y\}$. Let L is number of graylevels of the test picture, and the set of every gray value $\{0, 1, 2, ..., L-1\}$ can be denoted as O, in such a way that:

$$F(X,Y) \in O \quad \forall(x,y) \in \text{Image.} \tag{2.22}$$

Then, the normalized histogram will be

$$S = (t_0, t_1, ..., t_{L-1}). \tag{2.23}$$

For a bilevel thresholding case, Eq. (2.5) becomes:

$$S(T) = a_0(T_1) + a_1(T_2) \tag{2.24}$$

$$E(T) = \max_T \{S(T)\} \tag{2.25}$$

where $T = \{T_1, T_2, \ldots, T_L\}$ is the threshold value, $S = \{a_0, a_1, \ldots, a_{L-1}\}$ is the normalized histogram, and $E(T)$ is the optimal threshold. For an RGB image case, the above technique is separately implemented for R, G, and B threshold cases. Other information on the SE can be found in reference [19].

Advantages: This approach provides better image enhancement compared to other procedures and presents best values of the image quality measures than Otsu's approach.

Limitations: Implementation and identification of the maximized threshold in this technique are quite complex compared to Otsu's approach.

2.3.5 KAPUR'S APPROACH

Kapur's function has been primarily proposed for thresholding grayscale pictures using histogram's entropy [20]. This technique finds the optimal threshold by maximizing the entropy.

For a considered image, threshold vector $T = (t_0, t_1, \ldots, t_{L-1})$, Kapur's entropy is given as follows.

Let us consider a chosen dimension of the grayscale image with L graylevels (0 to $L-1$) with a total pixel value of Z. If $f(k)$ represents the frequency of kth intensity level, then the pixel distribution of the image will be:

$$Z = f(0) + f(1) + \cdots + f(L-1). \tag{2.26}$$

If the probability of kth intensity level is given by:

$$p(k) = f(k) / Z. \tag{2.27}$$

Then during the threshold selection, the pixels of image are separated into $T+1$ groups according to the assigned threshold value. After extrication of images as per the selected threshold, the entropy of each cluster is independently calculated and combined to get the final entropy as follows:

$$\text{Bilevel threshold} = f(t_1, t_2) = e_0 + e_1 \tag{2.28}$$

$$\text{Multilevel threshold} = f(t_1, t_2, \ldots t_L) = e_0 + e_1 + \cdots + e_{L-1} \tag{2.29}$$

$$e_0 = -\sum_{k=0}^{k=t_1-1} \frac{p_k}{\sigma_0} \ln \frac{p_k}{\sigma_0}, \sigma_0 = \sum_{k=0}^{k=t_1-1} p_k$$

$$e_1 = -\sum_{k=t_{1-1}}^{k=t_{1-2}} \frac{p_k}{\sigma_1} \ln \frac{p_k}{\sigma_1}, \sigma_1 = \sum_{k=t_{1-1}}^{k=t_{1-2}} p_k \qquad (2.30)$$

$$e_{L-1} = -\sum_{k=t_{L-1}}^{k=t_{L-2}} \frac{p_k}{\sigma_{L-1}} \ln \frac{p_k}{\sigma_{L-1}}, \sigma_{L-1} = \sum_{k=t_{L-1}}^{k=t_{L-2}} p_k$$

where e = entropy, p = probability distribution, and σ = probability occurrence.

$$\text{Kapur}_{\text{max}}(T) = \sum_{p=1}^{L-1} H_j^C. \qquad (2.31)$$

Other information on Kapur's function can be found in reference [21].

Advantages: It is one of the widely considered entropy function and provides better result on a class of grayscale and RGB images.

Limitations: The image quality measures attained with this technique are poor compared to that of SE.

2.4 SELECTION OF THRESHOLDING METHOD

From the above discussion, it is clear that thresholding process improves the texture and visibility of the image as per user need. The selection of right procedure (Otsu or entropy) and choice of the threshold value (bilevel or multilevel) need the following considerations:

- The type of image to be processed (color, dimension, histogram complexity, etc.): The literature confirms that Otsu's technique is one of the common procedures used to process a variety of images with lesser computational complexity. Next to the Otsu, Kapur's function is widely implemented. Kapur's technique is needed when assessment of the image abnormality is a primary task. Tsallis and Shannon's techniques are chosen to enhance RGB scale images compared to grayscale pictures. TE can be implemented where a bilevel thresholding is needed. Further, entropy methods can be used for the medical image examination tasks compared to Otsu's technique.
- Complexity in image: For simple image with even pixel distribution, Otsu's technique can be used, and for complex cases, a chosen entropy technique can be implemented.

The choice of threshold function depends mainly on the prior knowledge and, if essential, a detailed comparison betweenOtsu's technique and a chosen entropy technique can be performed to confirm the method to be used further. The comparison of Otsu and Kapur's techniques can be found in earlier works [7, 8].

2.5 IMPLEMENTATION ISSUES

Implementation of a bilevel thresholding is quite simple compared to multilevel thresholding. Also, thresholding of a grayscale image is straightforward compared to RGB. Hence, before implementing a chosen threshold operation, it is essential to verify the following constraints: dimension of the image, pixel distribution, color, and objective function to be satisfied [22–25].

- Color: The color of image pixel is the first choice, since the grayscale image has a single histogram and RGB has three histograms for each prime pixel (i.e., R, G, and B). The thresholding procedure developed for the RGB class will work on the grayscale picture. But, the approach developed for grayscale picture will not offer the result with the RGB image. Hence, it is necessary to choose appropriate procedure for RGB samples.
- Dimension: The complexity in the image enhancement also increases with increase in the dimension. If the dimension is more, the number of image pixels to be examined becomes large, which increases the thresholding burden. Hence, in most of medical image examination tasks, the image dimension is fixed as $256 \times 256 \times 1$ or $256 \times 256 \times 3$.
- Pixel distribution: Uneven pixel distribution and image associated with the noise also improve the complexity during the thresholding process. Hence, a pre-processing technique is used along with the thresholding process to reduce the complexity.

2.6 SUMMARY

This chapter presents the information on thresholding process and implementation of bilevel and multilevel thresholding processes on a chosen test picture. Further, details regarding well-known approaches, such as Otsu, Tsallis, Fuzzy-Tsallis, Shannon, and Kapur, are also presented. Also, choice and implementation issues in the image thresholding are discussed briefly.

REFERENCES

1. Rajinikanth, V., Satapathy, S.C., Fernandes, S.L., & Nachiappan, S. (2016). Entropy based segmentation of tumor from brain MR images–A study with teaching learning based optimization. *Pattern Recognition Letters*, 94, 87–94.
2. Rajinikanth, V., Dey, N., Kumar, R., Panneerselvam, J., & Raja, N.S.M. (2019). Fetal head periphery extraction from ultrasound image using Jayaalgorithm and Chan-Vese segmentation. *Procedia Computer Sciences*, 152, 66–73.
3. Dey, N., et al (2019). Social-group-optimization based tumor evaluation tool for clinical brain MRI of flair/diffusion-weighted modality. *Biocybernetics Biomedical Engineering*, 39(3), 843–856.
4. Fernandes, S.L., Rajinikanth, V., & Kadry, S. (2019). A hybrid framework to evaluate breast abnormality using infrared thermal images. *IEEE Consumer Electronics Magazine*, 8(5), 31–36.

5. Agrawal, S., Panda, R., Bhuyan, S., & Panigrahi, B.K. (2013). Tsallisentropy based optimal multilevel thresholding using cuckoo search algorithm. *Swarm and Evolutionary Computation*, 11, 16–30.

6. Ghamisi, P., Couceiro, M.S., & Benediktsson, J.A. (2013). Classification of hyperspectral images with binary fractional order Darwinian PSO and random forests. *SPIE Remote Sensing*, 8892, 88920S.

7. Lee, S.U., Chung, S.Y., & Park, R.H. (1990). A comparative performance study techniques for segmentation. *Computer Vision, Graphics, and Image Processing*, 52(2), 171–190.

8. Sezgin, M., & Sankar, B. (2004). Survey over image thresholding techniques and quantitative performance evaluation. *Journal of Electron Imaging*, 13(1), 146–165.

9. Lakshmi, V.S., Tebby, S.G., Shrirakjani, D., & Rajinikanth, V. (2016). Chaotic cuckoo search and Kapur/Tsallis approach in segmentation of T. Cruzi from blood smear images. *International Journal Computer Sciences and Information Security*, 14(CIC 2016), 51–56.

10. Rajinikanth, V., Satapathy, S.C., Dey, N., Fernandes, S.L., & Manic, K.S. (2019). Skin melanoma assessment using Kapur's entropy and level set—A study with bat algorithm. *Smart Innovation, Systemand Technologies*, 104, 193–202.

11. Rajinikanth, V., Dey, N., Raj, A.N.J., Hassanien, A.E., Santosh, K.C., & Raja, N.S.M. (2020). Harmony-search and Otsu based system for coronavirus disease (COVID-19) detection using lung CT scan images, *arXiv preprint*, arXiv:2004.03431.

12. Otsu, N. (1979). A threshold selection method from gray-level histograms. *IEEE Transactionsof System, Man, and Cybernetics*, 9(1), 62–66.

13. Raja, N.S.M., Rajinikanth, V., & Latha, K. (2014). Otsu based optimal multilevel image thresholding using firefly algorithm. *Modelling and Simulation in Engineering*, 2014(794574), 17.

14. Tsallis, C. (1988). Possible generalization of Boltzmann–Gibbs statistics. *Journal of Statistical Physics*, 52(1), 479–487.

15. Sadek, S., & Al-Hamadi, A. (2015). Entropic image segmentation: A fuzzy approach based on Tsallisentropy. *Internation Journal Computure Vision Signal Processessing*, 5(1), 1–7.

16. Sarkar, S., Paul, S., Burman, R., Das, S., & Chaudhuri, S.S. (2014) A fuzzy entropy based multilevel image thresholding using differential evolution. *Lecture Notes in Computer Science*, 8947, 386–395.

17. Anusuya, V., & Latha, P. (2014). A novel nature inspired fuzzy Tsallis entropy segmentation of magnetic resonance images. *Neuroquantology*, 12(2), 221–229.

18. Kannappan, P.L. (1972). On Shannon's entropy, directed divergence and inaccuracy. *Probability Theory and Related Fields*, 22, 95–100.

19. Paul, S., & Bandyopadhyay, B. (2014). A Novel Approach for Image Compression Based on MultiLevel Image Thresholding Using Shannon Entropy and Differential Evolution, In: *IEEE Students' Technology Symposium (TechSym)*, 56–61.

20. Kapur, J.N., Sahoo, P.K., & Wong, A.K.C. (1985). A new method for gray-level picture thresholding using the entropy of the histogram. *Computer Vision, Graphics, and Image Processing*, 29, 273–285.

21. Manic, K.S., Priya, R.K., & Rajinikanth, V. (2016). Image multithresholding based on Kapur/Tsallis entropy and firefly algorithm. *Indian Journal of Science Technology*, 9(12), 89949.

22. Akay, B. (2013). A study on particle swarm optimization and artificial bee colony algorithms for multilevel thresholding. *Applied Soft Computing Journal*, 13(6), 3066–3091.

23. Dougherty1994). *Digital Image Processing Methods*, 1stEdition, Boca Raton, FL: CRC Press.
24. Satapathy, S.C., Raja, N.S.M., Rajinikanth, V., Ashour, A.S., & Dey, N. (2018). Multilevel image thresholding using Otsu and Chaotic bat algorithm. *Neural Computing and Applications*, *29(12),*1285–1307.
25. Rajinikanth, V., & Couceiro, M.S. (2015). RGB histogram based color image segmentation using firefly algorithm. *Procedia Computer Science*, *46*, 1449–1457.

3 Grayscale and RGB-Scale Image Examination

3.1 IMAGE SELECTION

Image-based data registering is adopted in a variety of domains. To extract the information with better accuracy, it is necessary to implement a chosen image processing technique. Selection of right image for processing is essential, and the choice of a particular image depends on the following parameters [1–3]:

- **Dimension:** Image dimension is the key parameter to be fixed during image assessment task. Since the pixel distribution varies according to the size, it is essential to fix the dimension of grayscale/RGB images before assessment. When a chosen image is processed with an appropriate procedure, images with lesser dimension may need smaller processing time compared with images with larger dimensions. In most of the recent machine-learning ($256 \times 256 \times 3$) and deep-learning ($224 \times 224 \times 3$) cases, images with lesser dimension are preferred to minimize the computational burden. Hence, it is necessary to fix the dimension of the image for a chosen application as per the need.
- **Imagescale:** As discussed earlier, RGB-scale images are complex to process compared with the grayscale images, since in RGB-scale image, key pixels (R, G, and B) are to be examined separately, and in grayscale image, only the gray pixels are to be examined.
- **Enhancement needed:** After choosing an image with appropriate dimension and appropriate pixel scale, it is necessary to fix the enhancement procedure that is to be implemented to improve the information of the raw picture. In this work, the commonly used enhancement procedure called thresholding is discussed with appropriate examples.

3.2 GRAYSCALE AND RGB-SCALE IMAGE

In the literature, there exist a considerable number of image assessment methods executed on a class of images. The general images are to be evaluated in their present form, since the color conversion may degrade their information. But, in the medical domain, the existing images can be examined in their original form or after reducing their complexity [4–9].

To demonstrate the difference between the grayscale and RGB-scale image, this section considered the benchmark image with a pseudo name Mandrill. Figure 3.1 depicts the grayscale/RGB image of the Mandrill, and from this, it is clear that the

(a) Greyscale image   (b) RGB image

FIGURE 3.1   Sample test image (Mandrill).

image complexity of Figure 3.1(b) is quite high compared to that of Figure 3.1(a). The essential technical information (shape and texture) in both cases are same. If the texture or shape feature examination is the prime task, then we can choose either Figure 3.1(a) or (b) for assessment, and the outcome will be approximately the same. The choice of image color depends on the operation to be executed and computation effort we really need to put. If a computer algorithm is used to examine these two images, then the processing time required for the RGB class picture is more compared to alternative.

Advantage: Visual information available in the RGB-scale picture is large compared to the grayscale picture. But the texture and shape features are approximately the same.

Limitation: The processing time needed for the RGB-scale picture is quite large and this time will increase with the image dimension.

3.3   COMPLEXITY DUE TO IMAGE DIMENSION

The information as well as the pixel intensity of an image varies according to its dimension, but the shape and texture features remain the same. To demonstrate the change in complexity based on the chosen image, an experimental result attained on the Mandrill image is considered and its result is depicted in Figure 3.2 for different image cases [10, 11].

In this figure, the images with various pixel values, such as $256 \times 256 \times 1$, $256 \times 256 \times 3$, $512 \times 512 \times 3$, and $1024 \times 1024 \times 3$ are considered, and their variation in complexity is demonstrated with the help of their histograms.

Figure 3.2(a) depicts images with an assigned dimension and Figure 3.2(b) shows the histograms. The histogram normally represents the pixel distribution with respect to thresholds. In this work, for gray as well as RGB images, the threshold value is

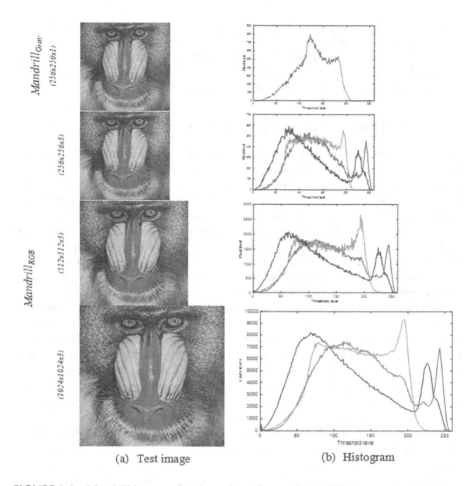

(a) Test image   (b) Histogram

FIGURE 3.2   Mandrill images of various dimensions and related histograms.

chosen as 256 (the span is 0–255), and from Figure 3.2(b), it can be noted that the histogram of grayscale image is quite simple compared to that of the RGB-scale picture. In the RGB-scale picture, the image consists of three different histograms to represent the prime pixel distributions, i.e., R, G, and B. In this work, a line plot attained with the MATLAB software is presented, and from this image, it can be seen that the green channel histogram approximately related with the pixel distribution of the grayscale image of the same dimension. But, the pixel density in the grayscale image is more compared to the RGB class.

When the dimension of RGB image increases, the pixel density will gradually rise by maintaining its distribution values (i.e., shape of the RGB histogram is approximately similar and it will not vary based on the size). Peaks and valleys in this threshold are to be examined during the threshold selection process. The identification of bilevel and multilevel thresholds is quite easy in grayscale case. For RGB class, every threshold (R, G, and B) is to be examined separately using a chosen thresholding approach. The combined value is then considered to examine the performance of implemented thresholding techniques.

From Figure 3.2, it can be noted that the complexity of thresholding process will rise based on the image dimension, and due to this, reduced images are widely preferred during automated image examination techniques.

3.4   COMPLEXITY DUE TO PIXEL DISTRIBUTION

Histogram-supported image examination procedures are performed to identify the finest threshold to enhance the image as per operator's requirement. During this process, an experienced operator or a chosen computer algorithm is allowed to adjust image thresholds till the requirement is satisfied. In this work, well-known threshold procedures, such as Otsu's and Kapur's approaches, are considered for demonstration. During this operation, the maximization of Otsu's between-class-variance or Kapur's entropy is chosen as the guiding criterion. If Otsu's thresholding is considered, then threshold values of these histograms are arbitrarily varied till Otsu's function is maximized. After maximizing this function, the thresholding operation stops and attained results are considered for further evaluation.

Varying the threshold value for the existing image histogram is quite complex, since the number of thresholds to be examined is 255. The combination and complexity of this process also increase with increase in the threshold value. Hence, a bounded threshold selection is recommended to minimize the computation effort. In the bounded search, the threshold of the image, which has lesser pixel level, is ignored/leftover from the search, and the bounded search may help to reduce the number of thresholds from 256 (0 to 255) to <256 (t_b to 255). The detailed discussion on the bounded search is discussed in references [9, 12, and 13].

The implementation of bounded search is demonstrated using the RGB histogram of the Mandrill image and is depicted in Figure 3.3. In the blue channel, the pixel distribution with respect to the threshold is very large (0 to 255) compared to the pixel distribution in green and red channels. For green and red channels, the exploration boundary is reduced by discarding the threshold with lesser pixel density. The threshold level considered in this work is represented with an arrow mark. From this, it can be understood that the complexity of thresholding related to the pixel distribution can be minimized with a bounded threshold search.

3.5   IMPLEMENTATION STEPS

During the image thresholding operation, the selection and implementation of an appropriate procedure are essential to enhance the image to a required level. In this work, the histogram-assisted image threshold is considered for demonstration. The implementation of a chosen procedure requires various steps as depicted below:

Step 1:   Select the test image of the chosen size
Step 2:   Implement image correction, if needed

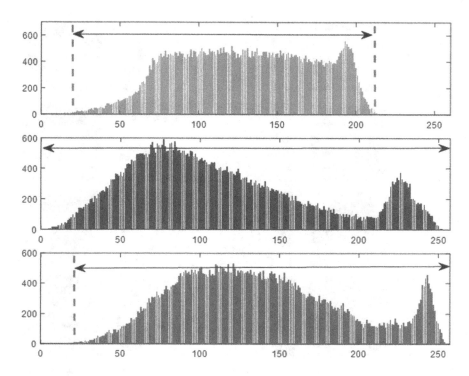

FIGURE 3.3 Bounded search with chosen pixel level.

Step 3: Choose and implement the objective function to be maximized

Step 4: Arbitrarily vary the threshold value and compute the objective function

Step 5: Compute the objective function for each combination of the threshold. If the objective value is maximized, stop the search and go to step 6 Else repeat steps 3 to 5

Step 6: Display the enhanced image and consider it for the assessment

Figure 3.4 demonstrates the thresholding outcome attained for the Mandrill image with Otsu and Kapur functions. In this work, initially, a bilevel threshold is implemented by choosing the threshold = 2 and the corresponding results are recorded for grayscale/RGB images. Later, a multilevel threshold with a chosen threshold of 3, 4, and 5 is implemented, and the corresponding results are recorded. After recording images, the performance of Otsu and Kapur functions is evaluated by computing the performance measures, such as Root Mean Squared Error, Peak Signal-to-Noise Ratio, Structural Similarity Index, Normalized Absolute Error, Normalized Cross Correlation, Average Difference, and Structural Content discussed in Section 1.6 of Chapter 1.

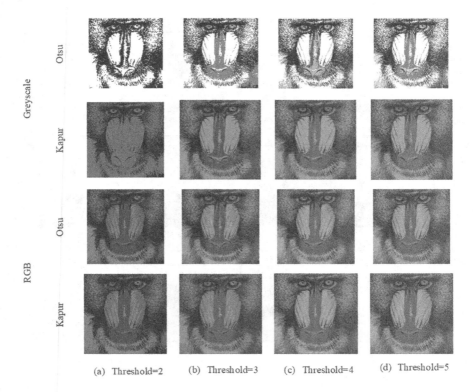

(a) Threshold=2   (b) Threshold=3   (c) Threshold=4   (d) Threshold=5

FIGURE 3.4   Bilevel and multilevel thresholding outputs attained for Mandrill image with Otsu and Kapur functions.

3.6 SUMMARY

This chapter presented the similarities and complexities between grayscale and RGB-scale images. To demonstrate this, a benchmark image with pseudo name Mandrill is considered. In this work, complexities in image due to their color space, dimension, and pixel distribution are clearly presented with the help of appropriate results. This work has also presented various steps that are considered during the thresholding process. The experimental outcomes of the Otsu and Kapur functions for bilevel and multilevel case are discussed. From this chapter, it can be noted that the complexity in the thresholding process can be minimized by choosing appropriate image and implementing the bounded threshold search.

REFERENCES

1. Lee, S.U., Chung, S.Y., & Park, R.H. (1990). A comparative performance study techniques for segmentation. *Computer Vision Graphics Image Processing*, 52(2), 171–190.
2. Sezgin, M., & Sankar, B. (2004). Survey over image thresholding techniques and quantitative performance evaluation. *Journal of Electron Imaging*, 13(1), 146–165.

3. Ghamisi, P., Couceiro, M.S., Benediktsson, J.A., & Ferreira, N.M.F. (2012). An efficient method for segmentation of images based on fractional calculus and natural selection. *Expert System with Applications*, 39(16), 12407–12417.

4. Sathya, P.D., & Kayalvizhi, R. (2011). Modified bacterial foraging algorithm based multilevel thresholding for image segmentation. *Engineering Applications of Artificial Intelligence*, 24, 595–615.

5. Abhinaya, B., & Raja, N.S.M. (2015). Solving multilevel image thresholding problem—An analysis with cuckoo search algorithm, information systems design and intelligent applications. *Advances in Intelligent Systems and Computing*, 339, 177–186.

6. Oliva, D., Cuevas, E., Pajares, G., Zaldivar, D., & Perez-Cisneros, M. (2013). Multilevel thresholding segmentation based on harmony search optimization. *Journal of Applied Mathematics*, 2013, 575414.

7. Ghamisi, P., Couceiro, M.S., Martins, F.M.L., & Benediktsson, J.A. (2014). Multilevel image segmentation based on fractional-order Darwinian particle swarm Optimization. *IEEE Transaction on Geoscience and Remote sensing*, 52(5), 2382–2394.

8. Manikantan,K., Arun, B.V., & Yaradonic, D.K.S. (2012). Optimal multilevel thresholds based on Tsalli sentropy method using golden ratio particle swarm optimization for improved image segmentation. *Procedia Engineering*, 30, 364–371.

9. Rajinikanth, V., Raja, N.S.M., & Latha, K. (2014). Optimal multilevel image thresholding: An analysis with PSO and BFO algorithms. *Australian Journal of Basic and Applied Sciences*, 8(9), 443–454.

10. Akay, B. (2013). A study on particle swarm optimization and artificial bee colony algorithms for multilevel thresholding. *Applied Soft Computing Journal*, 13(6), 3066–3091.

11. Satapathy, S.C., Raja, N.S.M., Rajinikanth, V., Ashour, A.S., & Dey, N. (2018). Multilevel image thresholding using Otsu and chaotic bat algorithm. *Neural Computing and Applications*, 29(12), 1285–1307.

12. Rajinikanth, V., & Couceiro, M.S. (2015). RGB histogram based color image segmentation using firefly algorithm. *Procedia Computer Science*, 46, 1449–1457.

13. Manic, K.S., Priya, R.K., & Rajinikanth, V. (2016). Image multithresholding based on Kapur/Tsallis entropy and firefly algorithm. *Indian Journal of Science and Technology*, 9(12), 89949.

4 Heuristic-Algorithm-Assisted Thresholding

4.1 THRESHOLDING METHODS

In the literature, a number of image thresholding procedures are proposed and executed on the RGB/grayscale images of various dimensions [1–3]. The identification of the threshold value for grayscale image is quite easy compared to the RGB-scale images. Since, in the RGB-scale image, the pixel distribution is quite complex and every pixel is mixture of R, G, and B pixels, the identification of the finest threshold is to be implemented individually for R, G, and B channels. Its complexity increases when the pixel level and distribution of pixels are nonlinear. In the case of grayscale, the pixel distribution is smooth and identification of the finest threshold is easy [4–9].

Let us consider a benchmark test image (Barbara) for the discussion and the dimensions of the test image are chosen as $720 \times 576 \times 3$ pixels (for RGB) and $720 \times 576 \times 1$ pixels (for grayscale) for demonstration.

Figure 4.1 presents the chosen benchmark image (Barbara) for experimental demonstration; Figure 4.1(a) and (c) present the test image and Figure 4.1(b) and (d) present corresponding histograms, respectively. From this figure, it can be noted that the histogram of RGB class is quite nonlinear and complex compared to the grayscale threshold.

In this work, histogram-based thresholding is considered, and the choice of the histogram is achieved by using a chosen methodology existing in the literature. During this search, the employed method is allowed to vary the thresholds of the image till the assigned objective function is maximized. The thresholding process can be implemented using the traditional (operator-assisted method) technique and computerized (heuristic-algorithm [HA]-assisted) technique. The choice mainly depends on the expertise of the operator and complexity of the thresholding process.

4.2 LIMITATIONS IN TRADITIONAL THRESHOLDING

Image threshold is one of the widely used image enhancement methods in which a chosen approach is used to enhance the test image based on the pixel grouping concept. When the separation to be implemented is bilevel (test image = ROT + background), then the identification of optimal threshold is quite simple and it can be implemented using the manual operator, if the image is in grayscale form. For the RGB-scale image and multilevel threshold selection, the computation time needed for a manual operator is quite high, and due to this complexity, HA-based techniques are widely implemented in recent times.

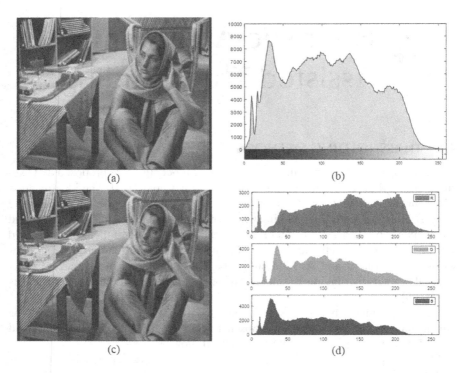

FIGURE 4.1   Barbara (gray/RGB) test image and related histograms.

For discussion, let us consider histogram of the Barbara image shown in Figure 4.2(a) and (b) and assume that the test image is to be separated into three pixel groups.

From Figure 4.2(a), it is clear that the gray pixel distribution is uniform from the values 0 to 255 and finding the required threshold value is quite complex. Further, the pixel closer to zero threshold is darker (background), and the pixel closer to the threshold 255 is light. The problem chosen in this study is to identify the threshold to find various pixel groups, such as Group1, Group2, and Group3, for both gray and RGB histograms.

If the grouping of pixel is carried out by a manual operator, a large number of threshold combinations are to be tried and the attained result is to be verified to confirm the superiority of the threshold operation. This procedure is tedious for RGB case and, hence, in recent years, HA-based techniques are widely employed to identify the optimal threshold for the test image by maximizing Otsu's function or a chosen entropy function.

HA-based procedures are recent image enhancement procedures, widely used in a variety of domains to preprocess and post-process a class of images using the most successful HAs existing in the literature. These algorithms considerably reduced the complexity in the image thresholding problem and work well on a class of images cases, even though it is assorted with various abnormalities.

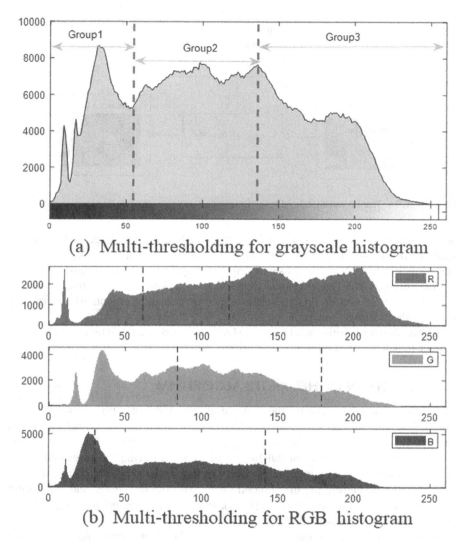

(a) Multi-thresholding for grayscale histogram

(b) Multi-thresholding for RGB histogram

FIGURE 4.2 Grouping of the image pixels during the multithreshold process.

4.3 NEED FOR HEURISTIC ALGORITHM

HAs are developed with the help of mathematical models of some well-known problem-solving capabilities found in the environment. Due to advancement in the computing technology, HAs can be easily implemented to solve a variety of constrained and unconstrained problems existing in the real world, and in this chapter, the chosen HA is utilized to solve the image thresholding problem.

Figure 4.3 presents a commonly existing histogram-based threshold problem, in which the histogram of the raw image is to be explored by the chosen HA in order to

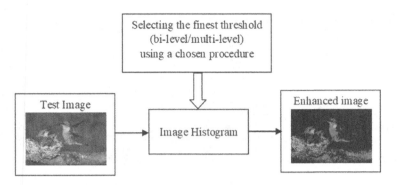

FIGURE 4.3   Traditional threshold selection process.

get the processed image. To demonstrate this, the benchmark image with a pseudo name Bird is considered. This image shows the output of a bilevel threshold operation in which the ROI is exactly separated with respect to the background.

Advantages: The HA technique works well on a class of images irrespective of its complexity. Moreover, this technique helped to reduce the computation burden of the multithresholding operation.

Limitations: Identification of a right HA is quite a challenging task and this technique needs few initial tunings for algorithm parameters.

4.4 SELECTION OF HEURISTIC ALGORITHM

The detailed arrangement of the HA-based threshold selection is depicted in Figure 4.4. The various stages involved in this operation include (i) selection of a suitable HA and tuning its parameters based on the problem to be solved, (ii) considering the objective-function (OF) to be maximized during the threshold selection, (iii) arbitrarily changing the histogram of the image till the maximized OF is attained, (iv) comparing the thresholded image with the test image and computing the essential image quality measures (IQM), and (v) validating the implemented procedure based on the attained IQM [10, 11].

From the above discussion, it is clear that the choice of a particular HA plays a major role in thresholding. In this chapter, well-known heuristic procedures such as particle swarm optimization (PSO), bacterial forging optimization (BFO), firefly algorithm (FA), bat algorithm (BA), cuckoo search (CS), and social group (SG) optimization are considered for the assessment.

4.4.1 PARTICLE SWARM OPTIMIZATION

PSO, proposed by Kennedy and Eberhart in 1995 [12], is a global optimization practice created with the motivation of community actions in bird and fish groups, and due to its performance, it is extensively functional in a variety of domains due to its elevated computational effectiveness. In contrast to other population-based

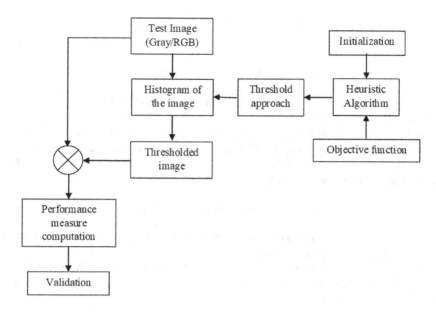

FIGURE 4.4 Multilevel threshold selection using heuristic algorithm.

stochastic techniques, PSO has equivalent or even better exploration performance for numerous hard optimization exertions, with faster and more stable convergence rates. It has been established to be a successful optimal means in various image processing domains [13, 14].

In PSO, the numbers of tuning parameters are few in comparison to other techniques. In this, a cluster of simulated birds is initialized with a random location L_i and speed S_i. At premature searching phase, every bird in the group is spread arbitrarily throughout the d sized exploration space. With the supervision of the OF, own flying knowledge, and their companions flying knowledge, every bird in the swarm energetically regulate its flying location and speed. During exploration, each particle remembers its best location reached so far (i.e., $p_{\text{best}} - (P_{i,d}^t)$) and also obtains the global best position information achieved by any particle in the population (i.e., $g_{\text{best}} - (G_{i,d}^t)$).

At iteration t, each particle i has its location defined by $L_{i,n}^t = [L_{i,1}, L_{i,2}, ..., L_{i,d}]$ and speed defined as $S_{i,n}^t = [S_{i,1}, S_{i,2}, ..., S_{i,d}]$ in search space d. Speed and location of each particle in the subsequent iteration can be calculated as:

$$S_{i,d}^{t+1} = W * S_{i,d}^t + q_1 * r_1 * (P_{i,d}^t - L_{i,d}^t) + q_2 * r_2 * (G_{i,d}^t - L_{i,d}^t) \qquad (4.1)$$

$$S_{i,d}^{t+1} = \Psi * \left[S_{i,d}^t + q_1 * r_1 * (P_{i,d}^t - L_{i,d}^t) + q_2 * r_2 * (G_{i,d}^t - L_{i,d}^t) \right] \qquad (4.2)$$

where $i = 1, 2, ..., k$ and $N = 1, 2, ..., d$.

$$L_{i,N}^{t+1} = \begin{cases} L_{i,N}^{t} + S_{i,N}^{t+1} & \text{if } L_{\min i,N} \leq L_i^{t+1} \leq L_{\max i,N} \\ L_{\min i,N} & \text{if } L_{i,N}^{t+1} < L_{\min i,N} \\ L_{\max i,N} & \text{if } L_{i,N}^{t+1} > L_{\max i,N} \end{cases} \qquad (4.3)$$

In Eqn. (4.1), the inertia of weight W represented is an important factor for PSO's convergence. It is used to control the impact of previous velocities on the current velocity. A large inertia weight factor facilitates global exploration, while small weight factor facilitates local exploration. Therefore, it is better to choose large weight factor for initial iterations and gradually reduce weight factor in successive iterations. This can be done by using the following equation:

$$w = w_{\max} - (w_{\max} - w_{\min}) * Iter / Iter_{\max} \qquad (4.4)$$

where $w_{\max}$ and $w_{\min}$ are initial and final weight, respectively, $Iter$ is iteration number, and $Iter_{\max}$ is the maximum iteration.

In Eqn. (4.2), constriction value Ψ is dependable for modernizing the speed of PSO which fix on the convergence and accuracy. This factor can be assigned using the following equation:

$$\Psi = \frac{2}{\left| 2 - \beta - \sqrt{\beta^2 - 4\beta} \right|}; \text{ where } \beta = q_1 + q_2, \beta > 4. \qquad (4.5)$$

Acceleration invariable q_1 called cognitive limit which pulls each element toward local best spot and constant q_2 called social limit pulls the particle near global best spot. r_1 and r_2 are known as arbitrary numbers in the range 0–1. The particle position is modified by Eq. (4.3). The process is repeated until the stopping criterion is reached. The flow chart of PSO is presented in Figure 4.5 [15, 16].

4.4.2 BACTERIAL FORAGING OPTIMIZATION

BFO is a type of biologically motivated stochastic exploring practice based on mimicking foraging (methods for locating, handling, and ingesting food) actions of *Escherichia coli* bacteria. Due to its qualities, such as efficiency, easy implementation, and stable convergence, it is widely applied to solve a range of complex engineering optimization problems [17–20].

In BFO, an OF is created as the endeavor for the bacteria in search of food. A set of simulated bacteria tries to achieve an optimum strength using phases: chemotaxis, swarming, reproduction, elimination, and dispersal. Every bacterium produces a result iteratively for a set of optimal values. Slowly all the microbes meet to the global optimum. In the chemotaxis stage, the bacteria either resort to a tumble, run, or swim. During swarming, every *E.coli* bacterium signals another bacterium such that attractants swarm mutually. In addition, during reproduction, the least healthy bacteria die and among the healthiest, each bacterium splits into two bacteria, which are placed at the same location. While in elimination and dispersal stages, any bacterium from the total set can be either eliminated or dispersed to a random location

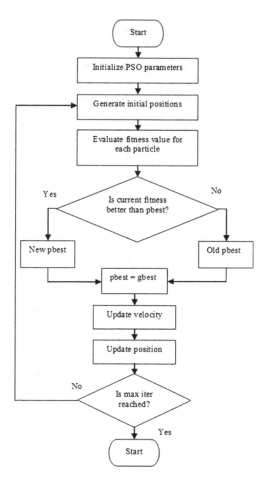

FIGURE 4.5 Flow chart of the PSO-based optimization.

during the optimization. This stage prevents the bacteria from attaining the local optimum. The flow chart of the BFO algorithm is depicted in Figure 4.6.

4.4.3 Firefly Algorithm

FA is one of the most successful HAs and the principal parameters that choose the effectiveness of the FA are the difference of light strength and attractiveness linking adjacent fireflies. These two parameters will be affected with an increase in the distance between fireflies and its introduction can be found in [21–23].

Change in luminance can be systematically articulated with the following Gaussian form:

$$I(r) = I_0 e^{-\gamma d^2} \tag{4.6}$$

where I = fresh brightness, I_0 = original brightness, and γ = light fascination coefficient.

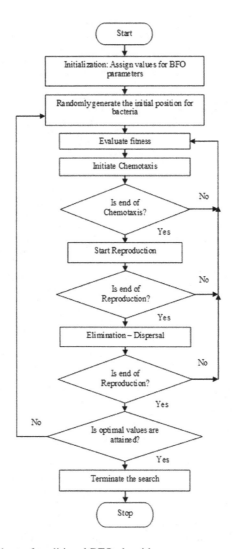

FIGURE 4.6 Flow chart of traditional BFO algorithm.

The attractiveness toward the luminance can be analytically represented as:

$$\gamma = \gamma_0 e^{-\beta d^2} \tag{4.7}$$

where γ = attractiveness coefficient and γ_0 = attractiveness at $r = 0$.

The above expression explain a quality space $\Gamma = 1/\sqrt{\beta}$ over which the attractiveness changes considerably from γ_0 to $\gamma_0 e - 1$. The attractiveness function $\gamma(d)$ can be any monotonically falling functions such as

$$\gamma(d) = \gamma_0 e^{-\beta d^q}, \ (q \geq 1). \tag{4.8}$$

For a fixed γ, the characteristic length becomes

$$\Gamma = \beta^{-1/q} \rightarrow 1, q \rightarrow \infty. \tag{4.9}$$

Conversely, for a given length scale Γ, the parameter γ can be used as atypical initial value (that is $\beta = 1/\Gamma q$).

The Cartesian space among two fireflies a and b at d_a and d_b in the G dimensional exploring space can be scientifically expressed as:

$$G_{ab}^t = \left\| d_b^t - d_a^t \right\|_2 = \sqrt{\sum_{k=1}^{n} (d_{b,k} - d_{a,k})^2}. \tag{4.10}$$

In FA, luminosity at a particular space G from the light source d_a^t obeys the inverse square law. The light intensity of a firefly I decreases, as the distance u increases in terms of $I \propto 1/u^2$. The movement of the attracted firefly a toward a brighter firefly b can be determined by the following position update equation:

$$d_a^{t+1} = d_a^t + \beta_0 e^{-\gamma u_{ab}^2} (d_b^t - d_a^t) + \Re \tag{4.11}$$

where d_a^{t+1} = updated position of firefly, d_a^t = initial position of firefly, $\beta_0 e^{-\gamma u_{ab}^2} (d_b^t - d_a^t)$ = attraction between fireflies, and $\Re$ = randomization parameter.

From Eqn. 4.11, it is observed that updated position of the ith firefly depends on initial position of the firefly, attractiveness of firefly toward the luminance, and the randomization parameter. Other information on FA can be found in [24–26].

4.4.4   BAT ALGORITHM

The BA was proposed in 2010 by Yang [27] to discover finest answer for arithmetic problems. Due to its worth, it was considered to solve various optimization problems. The process of conventional BA was connected with echolocation/biosonar character of microbats and it is arithmetically represented to construct the BA.

The BA has the following mathematical expressions:

$$\text{Velocity modernization} = S_i^{k+1} = S_i^k + [x_i^k - G_{best}]f_i \tag{4.12}$$

$$\text{Location update} = x_i^{k+1} = x_i^k + S_k^{k+1} \tag{4.13}$$

$$\text{Frequency alteration} = f_i = f_{min} + (f_{max} - f_{min})\alpha \tag{4.14}$$

where α is an arbitrary value of choice [0, 1].

Eqn. (4.14) controls Eqn. (4.12) and Eqn. (4.13), and hence, the choice of the frequency value should be appropriate.

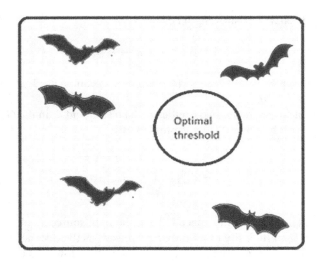

FIGURE 4.7 Optimal threshold identification with bat algorithmm.

Updated value for every bat is produced based on:

$$x_{new} = x_{old} + \varepsilon A^k \tag{4.15}$$

where ε is an arbitrary value of choice [−1, 1] and A = loudness constraint during the exploration.

The expression of the loudness variation can be represented as:

$$A_i^{n+1} = \varphi A_i(k) \tag{4.16}$$

where φ is a variable with a value $0 < \varphi < 1$.

Other values on the BA can be found in the literature [28, 29].

The conventional working of the BA is shown in Figure 4.7.

The objective of the BA is to recognize the optimal answer for a specified problem by exploring the search area. If a single bat identifies the optimal solution, then it will invite other bats toward the solution. The probability of getting the global maxima in BA is better compared to other algorithms, and hence, it is one of the successful approaches and used in a variety of optimization tasks.

4.4.5 CUCKOO SEARCH

The traditional CS was invented by Yang and Deb in 2009 [30], and over a decade, it is extensively used to resolve an assortment of image processing applications. In the literature, a considerable number of improvement procedures are planned to develop the optimization exploration of the CS.

The traditional CS is developed by mimicking the breeding artifice followed by parasitical cuckoos. The CS is developed by considering the following assumptions:

(i) every cuckoo deposit an egg in a randomly selected nest of the host birds, (ii) the nest with eminent enduring egg will be carried to the subsequent invention, and (iii) for a chosen threshold problem, the amount of host bird's nest is fixed and it may identify the egg of cuckoo with a probability $P_a \in [0,1]$. When the host identifies the egg, it may eliminate the egg or abandon the current nest and the host will build a new nest.

Above-said assumptions are accounted to create the mathematical model of the CS. The implementation steps in CS are quite simple compared to other existing approaches and the proposed CS is depicted in Eqn. (4.17)

$$P_i^{n+1} = P_i^n + \sigma \oplus LF \tag{4.17}$$

where P_i^n = early position, P_i^{n+1} = updated position, σ = step size (chosen as 1.2), $\oplus$ = entry wise multiplication, and LF = Levy-Flight operator.

The pseudo-code of the CS is depicted below and its complete detail can be found in references [28, 31]:

Initialize the CS with; $N = no.\ of\ nests$, Iter_{max} = maximum iterations, $P_n \in [0,1]$, $F(P) = Objective$ function, **and** *Stopping* constraint $(F_{max}(P)$ or $\text{Iter}_{max})$		
Initialize the counter (Set n = 0)		
for (i=1: I ≤ N) do		
		Initiate the population of N-host P_i^n
		Evaluate $F(P_i^n)$
end for		
repeat		
		Generate X_i^{n+1} using Eqn. (4.17)
		Evaluate $F(P_i^{n+1})$
		Chose a nest P_j randomly from N-solutions
if $\{F(P_i^{n+1})\} > \{F(P_j^n)\}$ **then**		
		Replace P_j with P_i^{n+1}
end if		
		Abandon he nest based on $P_a \in [0,1]$
		Built a new nest
		Keep the best solutions (nest with best solutions)
		Rank and sort the solutions to find the best one
Increase iteration count (Set n = n + 1)		
repeat till *stopping constraint reached* $(F_{max}(P)$ or $Iter_{max})$		
Produce the optimized result		

4.4.6 SOCIAL GROUP OPTIMIZATION

SGO is an HA developed by Satapathy and Naik [32]. It has been created by imitating the performance and knowledge conveying followed in human groups. The SG includes two chief functions, namely, (i) improving step, which synchronizes the location of citizens (agents) based on the OF, and (ii) acquiring step that permits the agents to find out the finest potential answer for the problem under concern.

The arithmetical replica for the SGO is as follows [33–35].

Let's consider K_i as the initial knowledge of people in a group and $i = 1, 2, 3,..., P$, with P as the total number of people in the group. If the optimization task needs a D-dimensional search space, then the knowledge term can be expressed as $K_i = (k_{i1}, k_{i2}, k_{i3}, ..., k_{id})$. For any problem, the fitness value can be defined as f_j, with $j = 1, 2, ..., P$. Thus, for the maximization problem, the fitness value can be written as

$$Gbest_j = max\{f(K_i) \ for \ i = 1, 2, ...,P\}. \qquad (4.18)$$

The steps of the standard SG algorithm can be described as follows:

Standard SG Optimization Algorithm
Start
Assume five agents ($i = 1,2,3,4,5$)
Assign these agents to discover the $Gbest_j$ in a D-dimensional exploration space
Randomly dispense the whole agents in the cluster throughout the exploration space throughout initialization practice
Compute the fitness cost based on the problem under apprehension
Update the orientation of agents with $Gbest_j = max \{f(K_i)\}$
Initiate the improving phase to modernize the knowledge of additional agents in order to attain the $Gbest_j$
Initiate the acquiring phase to advance the knowledge of agents by arbitrarily choosing the agents with best fitness
Repeat the procedure till the entire agents move toward the best possible position in the D-dimensional exploration space
If all the agents have approximately alike strength values ($Gbest_j$)
  Then
Terminate the explore and display the optimized result for the chosen problem
  Else
  Repeat the previous steps
End
Stop

In order to update the position (knowledge) of every individual in the group, the improving phase considers the following relation:

$$K_{new_{i,j}} = c * K_{old_{i,j}} + R * (Gbest_j - K_{old_{i,j}}) \qquad (4.19)$$

where K_{new} = new knowledge, K_{old} = old knowledge, $Gbest$ = global best knowledge, R = random numeral [0, 1], and c represents the self-introspection parameter [0, 1] and its value is chosen as 0.2.

During the acquiring phase, the agents will find the global solution based on the knowledge updating process by randomly selecting one person from the group (K_r) based on $i \neq r$. Once the fitness value becomes $f(K_i) < f(K_r)$, then the following knowledge procedure is executed:

$$K_{new_{i,j}} = K_{old_{i,j}} + R_a * (K_{i,j} - K_{r,j}) + R_b * (Gbest_j - K_{i,j}) \qquad (4.20)$$

where R_a and R_b are random numbers having the range [0, 1] and $K_{r,j}$ is the knowledge (position) value of chosen individual.

4.5 IMPLEMENTATION STEPS

Implementation of the HA-assisted threshold selection is clearly described in Figure 4.4. After selecting the appropriate HA, it is necessary to initialize a tuning procedure for the algorithm, which includes the selection of number of agents, fixing the dimension of search (search dimension = number of thresholds to be identified), OF to be maximized, stopping criterion and validation of the performance of the proposed thresholding experiment. Normally, a histogram-based thresholding is implemented with HA, in which the image thresholds are randomly varied till the optimal threshold is achieved. The technique implemented with the HA is quite simple and it helps to attain better and faster result compared to the traditional thresholding approach.

4.6 SUMMARY

This chapter clearly presented grayscale/RGB threshold methods based on traditional and HA-assisted techniques. Further, a detailed graphical description is presented to describe the implementation of HA-based multithresholding process. Furthermore, the outline of some famous HAs such as PSO, BFO, FA, BA, CS, and SGO are presented.

REFERENCES

1. Oliva, D., Cuevas, E., Pajares, G., Zaldivar, D., & Perez-Cisneros, M. (2013). Multilevel thresholding segmentation based on harmony search optimization. *Journal of Applied Mathematics*, 2013, 575414.
2. Ghamisi, P., Couceiro, M.S., Martins, F.M.L., & Benediktsson, J.A. (2014). Multilevel image segmentation based on fractional-order Darwinian particle swarm optimization. *IEEE Transaction on Geoscience and Remote Sensing*, 52(5), 2382–2394.
3. Satapathy, S.C., Raja, N.S.M., Rajinikanth, V., Ashour, A.S., & Dey, N. (2018). Multilevel image thresholding using Otsu and chaotic bat algorithm. *Neural Computing and Applications*, 29(12), 1285–1307.

4. Rajinikanth, V., Satapathy, S.C., Fernandes, S.L., & Nachiappan, S. (2016). Entropy based segmentation of tumor from brain MR images–A study with teaching learning-based optimization. *Pattern Recognition Letters*, 94, 87–94.

5. Rajinikanth, V., & Couceiro, M.S. (2015). RGB histogram based color image segmentation using firefly algorithm. *Procedia Computer Science*, 46, 1449–1457.

6. Manic, K.S., Priya, R.K., & Rajinikanth, V. (2016). Image multi thresholding based on Kapur/Tsallis entropy and firefly algorithm. *Indian Journal of Sciences Technolonogy*, 9(12), 89949.

7. Sathya, P.D., & Kayalvizhi, R. (2011). Modified bacterial foraging algorithm based multilevel thresholding for image segmentation. *Engineering Applications of Artificial Intelligence*, 24, 595–615.

8. Abhinaya, B., & Raja, N.S.M.. (2015). Solving multilevel image thresholding problem—Ananalysis with cuckoo search algorithm, information systems design and intelligent applications. *Advances in Intelligent Systems and Computing*, 339, 177–186.

9. Akay, B. (2013). A study on particle swarm optimization and artificial bee colony algorithms for multilevel thresholding. *Applied Soft Computing Journal*, 13(6), 3066–3091.

10. Rajinikanth, V., Raja, N.S.M., & Kamalanand, K. (2017). Firefly algorithm assisted segmentation of tumor from brain MRI using Tsallis function and Markov random field. *Journal of Control Engineering and Applied Informatics*, 19(3), 97–106.

11. Palani, T.K., Parvathavarthini, B., & Chitra, K. (2016). Segmentation of brain regions by integrating meta heuristic multilevel threshold with Markov random field. *Current Medical Imaging Review*, 12(1), 4–12.

12. Eberhart, R.C., & Shi, Y. (2001). Tracking and Optimizing Dynamic Systems With Particle Swarms. In: *Proceedings of the IEEE Congress on Evolutionary Computation*, 94–100. IEEE, Seoul, Korea.

13. Eberhart, R.C., Simpson, P.K., & Dobbins, R.W. (1996). *Computational Intelligence PC Tools*. Academic Press, Boston, MA.

14. Blackwell, T.M. (2005). Particle swarms and population diversity. *Soft Computing*, 9, 793–802.

15. Clerc, M. (2006). *Particle Swarm Optimization*. ISTE, London, U.K.

16. Iwasaki, N., & Yasuda, K.(2005). Adaptive particle swarm optimization using velocity feedback. *International Journal of Innovative Computing, Information and Control*, 1(3), 369–380.

17. Passino, K.M.. (2002). Biomimicry of bacterial foraging for distributed optimization and control. *IEEE Control Systems Magazine*, 22(3), 52–67.

18. Das, S., Biswas, A., Dasgupta, S., & Abraham, A. (2009). Bacterial Foraging Optimization Algorithm: Theoretical Foundations, Analysis, and Applications. In: Abraham, A, Hassanien, AE, Siarry, P, & Engelbrecht, A. (eds.), *Foundations of Computational Intelligence Volume 3. Studies in Computational Intelligence*, vol 203. Springer–Berlin, Germany.

19. Liu, Y., & Passino, K.M. (2002). Biomimicry of social for aging bacteria for distributed optimization: Models, principles, and emergent behaviors. *Journal of Optimization Theory and Applications*, 115(3), 603–628.

20. Abraham, A., Biswas, A., Dasgupta, S., & Das, S.(2008). Anaysis of Reproduction Operator in Bacterial Foraging Optimization. In: *IEEE World Congress on Computational Intelligence, WCCI 2008*, 1476–1483. IEEE Press, USA.

21. Yang, X.S. (2009). Firefly Algorithms for Multimodal Optimization. *In: Proceeding of the Conference on Stochastic Algorithms: Foundations and Applications*, 169–178, Sapporo, Japan.

22. Yang, X.S. (2010). Firefly Algorithm, Levy Flights and Global Optimization. *In: Watanabe, O., Zeugmann, T. (eds.), Research and Development in Intelligent Systems XXVI*, 209–218. Springer, Berlin, Germany.

23. Yang, X.S. (2013). Multiobjective firefly algorithm for continuous optimization. *Engineering Computers* 29, 175–184
24. Yang, X.S. (2010). Firefly algorithm, stochastic test functions and design optimisation. *International Journal of Bio-Inspired Computation*, 2(2), 78–84.
25. Yang, X.S. (2011). Review of meta-heuristics and generalised evolutionary walk algorithm. *International Journal of Bio-Inspired Computation*, 3(2), 77–84.
26. Dey, N., (2020). Applications of Firefly Algorithm and Its Variants. *Springer Tracts in Nature-Inspired Computing*. Springer, Berlin, Germany.
27. Yang, X.-S. (2010). A New Meta heuristic Bat-Inspired Algorithm. *In: González, J.R., Pelta, D.A., Cruz, C., Terrazas, G., Krasnogor, N.(eds.) NICSO 2010*. SCI, 284, 65–74. Springer, Berlin, Germany.
28. Yang, X.-S. (2010) *Nature-Inspired Metaheuristic Algorithms*. Luniver Press, UK.
29. Dey, N., & Rajinikanth, V. (2020). Applications of Bat Algorithm and Its Variants. In: *Springer Tracts in Nature-Inspired Computing*. Springer, Berlin, Germany.
30. Yang, X.-S., & Deb, S.. (2009) Cuckoo search via Lévy flights. In: *World Congress on Nature & Biologically Inspired Computing*. Coimbatore, India.
31. Burnwal, S., & Deb, S. (2013). Scheduling optimization of flexible manufacturing system using cuckoo search-based approach. *International Journal of Advanced Manufacturing Technology*, 64(5–8), 951–959.
32. Satapathy, S., & Naik, A. (2016). Social group optimization (SGO): A new population evolutionary optimization technique. *Complex & Intelligent Systems*, 2(3), 173–203.
33. Naik, A., Satapathy, S.C., Ashour, A.S., & Dey, N. (2016). Social group optimization for global optimization of multimodal functions and data clustering problems. *Neural Computing and Application*, 30, 271–287. Available online: https://doi.org/10.1007/s00521-016-2686-9.
34. Rajinikanth, V., & Satapathy, S.C. (2018). Segmentation of ischemic stroke lesion in brain MRI based on social group optimization and fuzzy-tsallis entropy. *Arabian Journal for Sciences and Engineering*, 43, 4365–4378. Available online: https://doi.org/10.1007/s13369-017-3053-6.
35. Dey, N., et al (2019). Social-group-optimization based tumor evaluation tool for clinical brain MRI of Flair/diffusion-weighted modality. *Biocybernetics Biomedical Engineering*, 39(3), 843–856.

5 Objective Function and Image Quality Measures

5.1 INTRODUCTION TO IMPLEMENTATION

The implementation of a chosen thresholding process is clearly depicted in Figure 4.4 and this procedure will work on a class of conventional and medical images [1-7]. The chief aim of this technique is to employ the operation on a chosen image to identify the finest threshold to group the image pixels. During implementation, the following procedures are to be considered:

Step 1: Collect the image to be examined and improve the image quality and dimension as needed.
(This procedure helps to fix constraints in the image and provides a good quality test image)
Step 2: Select the appropriate OF and HA to achieve the optimal result.
(This procedure helps to choose Otsu-/Entropy-based technique to implement thresholding)
Step 3: Fix the number of threshold needed to enhance the ROI of the test image for further assessment.
(The threshold value will be assigned as 2 for bilevel operation and threshold >2 for the multilevel operation)
Step 4: Execute the threshold procedure and repeat it over a predefined times. Apply the statistical test to confirm robustness of the proposed technique.
Step 5: After attaining the result, compute the IQM.
Step 6: Confirm the superiority of the implemented technique and validate the results.

5.2 MONITORING PARAMETER

Normally, the HA-based multilevel thresholding is an automated technique performed using dedicated computer algorithms. Due to its practical significance, recently a number of image processing application adopted HA-assisted techniques [8–10]. The quality and quantity of the outcome depend mainly on monitoring functions employed to supervise the automated operation. In most of the cases, a carefully assigned OF is used as a supervisor, which controls the entire image thresholding operation. Due to this reason, it is essential to have a superior objective function to monitor and control the computerized algorithms [11–14].

The choice of OF depends mainly on the procedure implemented to perform the operation. If Otsu's function is implemented, the maximized value of the between-class-variance act as the OF, and in the case of Tsallis, FT, Shannon, and Kapur, the maximized entropy act as the OF. According to the requirement, we can employ a single OF or multiple OF. The upcoming section outline the implantation of the OF for a chosen method [15, 16].

5.2.1 OBJECTIVE FUNCTION

OF plays a foremost role in the image thresholding and its choice depends based on the implemented technique.

Some commonly used OFs along with their procedure are depicted below:

- Otsu's between-class variance: If this function is implemented, the OF will be the maximization of Otsu's constraint depicted in Eqn 5.1

$$\text{Otsu}_{\max} = J(T) = \vartheta_0 + \vartheta_1 + \cdots + \vartheta_{L-1} \tag{5.1}$$

  where $\vartheta_0 = \eta_0(\psi_0 - \psi_T)^2$, $\vartheta_1 = \eta_1(\psi_1 - \psi_T)^2$, ..., $\vartheta_T = \eta_T(\psi_T - \psi_{L-1})^2$.
- Tsallis entropy: In this, the entropy of the histogram is considered as the OF, and the mathematical expression is depicted in Eqn. 5.2

$$\text{Tsallis}(t_i) = [t_0, t_1, ..., t_{L-1}] = \arg\max[S_{hQ}^{F_1}(t) + S_Q^{F_2}(t) + \cdots + S_Q^{k}(t) \\ + (1-Q) \cdot S_Q^{F_1}(t) \cdot S_Q^{F_2}(t), ..., S_Q^{F_K}(t)]. \tag{5.2}$$

- FT entropy: In this, the entropy of the histogram is considered as the OF, and the mathematical expression is depicted in Eqn. 5.3

$$\text{Tsallis}_{\max}(T) = \arg\max[S_Q^1(T) + S_Q^2(T)(1-Q) \cdot S_Q^1(T) \cdot S_Q^2(T)]. \tag{5.3}$$

- Shannon's entropy: In this, the entropy of the histogram is considered as the OF, and the mathematical expression is depicted in Eqn. 5.4

$$E(T) = \max_T\{S(T)\}. \tag{5.4}$$

- Kapur's entropy: The entropy of the histogram is considered as the OF, and the mathematical expression is depicted in Eqn. 5.5

$$\text{Kapur}_{\max}(T) = \sum_{p=1}^{L-1} H_j^C. \tag{5.5}$$

5.2.2 SINGLE AND MULTIPLE OBJECTIVE FUNCTION

The quality of the thresholded image depends on OF and based on the requirement, we can implement single or multiple objective functions.

The expressions depicted in Eqn. (5.1)–(5.5) present the information on single OF. The multiple OF is the modified version of the single OF, and for the demonstration, this section considers Otsu's function and similar technique can be used for other entropy functions

$$OF_{single} = Otsu_{max} = J(T) = \vartheta_0 + \vartheta_1 + \cdots + \vartheta_{L-1} \tag{5.6}$$

$$OF_{multiple1} = (W_1 * Otsu_{max}) + (W_2 * PSNR) \tag{5.7}$$

$$OF_{multiple2} = (W_1 * Otsu_{max}) + (W_2 * PSNR) + (W_3 * SSIM) \tag{5.8}$$

where W_1, W_2, and W_3 are weighting functions, whose values are assigned as [0,1]. Eqn. (5.6) depicts single OF, and Eqns. (5.7) and (5.8) presents multiple OF.

5.3   ASSESSMENT OF THRESHOLDING PROCESS

The superiority of the threshold outcome can be assessed based on a comparative analysis between raw image (R) and thresholded image (T). During this operation, each pixel of these images is separately compared, and based on its value, the essential image quality parameters are computed. The quality of the T is assessed with well-known image metrics, such as Root Mean Squared Error (RMSE), Peak Signal-to-Noise Ratio (PSNR), Mean Structural Similarity Index (MSSIM), Normalized Absolute Error (NAE), Normalized Cross-Correlation (NCC), Average Difference (AD), and Structural Content (SC) [17–20].

Mathematical expression of the considered image quality measures are presented as:

$$PSNR_{(R,T)} = 20 \log_{10} \left(\frac{255}{\sqrt{MSE_{(R,T)}}} \right); dB \tag{5.9}$$

$$RMSE_{(R,T)} = \sqrt{MSE_{(R,T)}} = \sqrt{\frac{1}{XY} \sum_{i=1}^{X} \sum_{j=1}^{Y} [R_{(i,j)} - T_{(i,j)}]^2}. \tag{5.10}$$

The mean SSIM is generally used to estimate the image superiority and inter dependencies between the original and processed image

$$MSSIM_{(R,T)} = \frac{1}{M} \sum_{z=1}^{M} SSIM_{(Rz,Tz)} \tag{5.11}$$

where R_z and T_z are the image contents at the z-th local window; and M is the number of local windows in the image

$$NAE_{(R,T)} = \frac{\sum_{i=1}^{X}\sum_{j=1}^{Y}\left|R_{(i,j)} - T_{(i,j)}\right|}{\sum_{i=1}^{X}\sum_{j=1}^{Y}\left|R_{(i,j)}\right|} \tag{5.12}$$

$$NCC_{(R,T)} = \frac{\sum_{i=1}^{X}\sum_{j=1}^{Y}R_{(i,j)} \cdot T_{(i,j)}}{\sum_{i=1}^{X}\sum_{j=1}^{Y}R^2_{(i,j)}} \tag{5.13}$$

$$AD_{(R,T)} = \frac{\sum_{i=1}^{X}\sum_{j=1}^{Y}R_{(i,j)} - T_{(i,j)}}{XY} \tag{5.14}$$

$$SC_{(R,T)} = \frac{\sum_{i=1}^{X}\sum_{j=1}^{Y}R^2_{(i,j)}}{\sum_{i=1}^{X}\sum_{j=1}^{Y}T^2_{(i,j)}}. \tag{5.15}$$

In all the expressions, $X * Y$ is the size of considered image, R is the original test image, and S is the segmented image of a chosen threshold. A higher value of PSNR, MSSIM, NCC and lower value of RMSE, NAE, AD, SC specify a superior quality of thresholding. Improved Tsallis fitness function with minor CPU time during the optimization search also confirms the capability of the considered heuristic algorithm.

In order to demonstrate the computation of IQV for the thresholded image, a multilevel thresholding is implemented and the attained results are presented in Figure 5.1. Figure 5.1 (a) and (b) depicts the considered benchmark test image and corresponding histogram, respectively. Figure 5.1 (c)–(f) presents Otsu's thresholding results attained for a threshold value of Th = 2,3,4, and 5, respectively. Further, Kapur's thresholding outcome is depicted in Figure 5.1 (g)–(j) for an assigned threshold value of Th = 2,3,4, and 5, respectively.

Table 5.1 depicts the results attained with the proposed technique for Otsu and Kapur. In this table, OF is the maximized OF and OT refers to optimal thresholds. Other values depict the IQV computed by comparing raw and thresholded images. The IQV attained with Otsu's function is superior compared to the valued attained with Kapur's function.

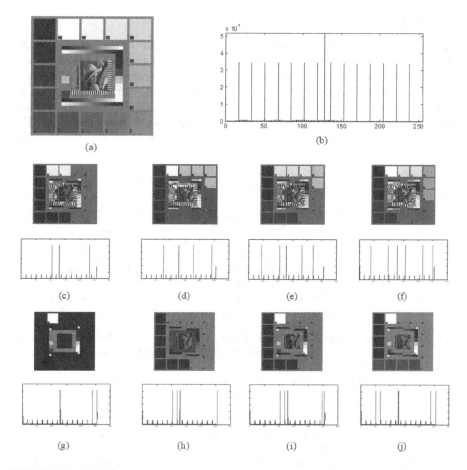

FIGURE 5.1 Multilevel thresholding achieved with Otsu's and Kapur's functions.

TABLE 5.1
Experimental Results Attained with the Multilevel Thresholding with Otsu and Kapur Functions

Method	Th	OF	OT	RMSE	PSNR	NAE	NCC	SSIM	CPU
Otsu	2	3662.82	105,233	47.7593	14.5496	0.3064	0.7479	0.6824	10.3769
	3	5213.74	77,179,242	44.2516	15.2122	0.3001	0.7630	0.7329	11.8867
	4	5414.60	44,106,182,220	28.2878	19.0988	0.1865	0.8513	0.8593	13.4487
	5	5574.41	40,99,158,216,251	26.8298	19.5584	0.1791	0.8629	0.8677	16.4567
Kapur	2	7.7831	239,255	122.1902	6.3901	0.8003	0.2981	0.1867	10.6285
	3	10.7531	103,119,256	65.3641	11.8240	0.3750	0.6171	0.5686	13.2928
	4	13.2309	103,119,246,253	49.8238	14.1821	0.2708	0.7726	0.6510	16.1187
	5	16.2953	52,68,128,239,255	38.0448	16.5249	0.1828	0.8303	0.7655	17.0026

5.4   SUMMARY

This chapter presented the selection of single and multiple OF for a chosen image. Further, this work also presents the information on various IQM to be computed to confirm the superiority of the threshold process. Comparison of the Otsu- and Kapur-based multilevel thresholding is experimentally demonstrated using a benchmark test image, and the results are presented. These results confirm that the thresholded outcome of Otsu's technique is superior compared to Kapur's technique. The implementation of the histogram-based multilevel threshold is also demonstrated with appropriate results.

REFERENCES

1. Rajinikanth, V., Satapathy, S.C., Fernandes, S.L., & Nachiappan, S. (2016). Entropy based segmentation of tumor from brain MR Images–A study with teaching learning based optimization. *Pattern Recognition Letters*, 94, 87–94.
2. Rajinikanth, V., Dey, N., Kumar, R., Panneerselvam, J., & Raja, N.S.M. (2019). Fetal head periphery extraction from ultrasound image using jaya algorithm and chan-vese segmentation. *Procedia Computer Science*, 152, 66–73.
3. Dey, N., et al (2019). Social-group-optimization based tumor evaluation tool for clinical brain MRI of Flair/diffusion-weighted modality. *Biocybernetics and Biomedical Engineering*, 39(3), 843–856.
4. Fernandes, S.L., Rajinikanth, V., & Kadry, S. (2019). A hybrid framework to evaluate breast abnormality using infrared thermal images. *IEEE Consumer Electronics Magazine*, 8(5), 31–36.
5. Agrawal, S., Panda, R., Bhuyan, S., & Panigrahi, B.K. (2013). Tsallis entropy based optimal multilevel thresholding using cuckoo search algorithm. *Swarm and Evolutionary Computation*, 11, 16–30.
6. Ghamisi, P., Couceiro, M.S., & Benediktsson, J.A. (2013). Classification of hyperspectral images with binary fractional order Darwinian PSO and random forests. *SPIE Remote Sensing*, 8892, 88920S.
7. Sathya, P.D., & Kayalvizhi, R. (2011). Modified bacterial foraging algorithm based multilevel thresholding for image segmentation. *Engineering Applications of Artificial Intelligence*, 24, 595–615.
8. Abhinaya, B., & Raja, N.S.M.. (2015). Solving multilevel image thresholding problem— Ananalysis with cuckoo search algorithm, information systems design and intelligent applications. *Advances in Intelligent Systems and Computing*, 339, 177–186.
9. Akay, B. (2013). A study on particle swarm optimization and artificial bee colony algorithms for multilevel thresholding. *Applied Soft Computing Journal*, 13(6), 3066–3091.
10. Rajinikanth, V., Raja, N.S.M., & Kamalanand, K. (2017). Firefly algorithm assisted segmentation of tumor from brain MRI using tsallis function and markov random field. *Journal of Control Engineering Applied Informatics*, 19(3), 97–106.
11. Lee, S.U., Chung, S.Y., & Park,R.H. (1990). A comparative performance study techniques for segmentation. *Comput Vision Graphics Image Processing*, 52(2), 171–190.
12. Sezgin, M., & Sankar, B. (2004). Survey over image thresholding techniques and quantitative performance evaluation. *Journal of Electron Imaging*, 13(1), 146–165.
13. Lakshmi, V.S., Tebby, S.G., Shriranjani, D., & Rajinikanth, V. (2016). Chaotic cuckoo search and Kapur/Tsallis approach in segmentation of T. Cruzi From blood smear images. *International Journal of Computer Science Information Security*, 14(CIC 2016), 51–56.

14. Rajinikanth, V., & Couceiro, M.S. (2015). RGB histogram based color image segmentation using firefly algorithm. *Procedia Computer Science*, *46*, 1449–1457.
15. Manic, K.S., Priya, R.K., & Rajinikanth, V. (2016). Image multithresholding based on Kapur/Tsallis entropy and firefly algorithm. *Indian Journal of Science and Technology*, 9(12), 89949.
16. Sarkar, S., Paul, S., Burman, R., Das, S., & Chaudhuri, S.S. (2014). A fuzzy entropy based multilevel image thresholding using differential evolution. *Lecture Notes in Computer Science*, 8947, 386–395.
17. Hore, A., & Ziou, D. (2010). Image Quality Metrics: PSNR vs. SSIM, In: *IEEE International Conference on Pattern Recognition*, 2366–2369, Istanbul, Turkey.
18. Wang, Z., Bovik, A.C., Sheikh, H.R., & Simoncelli, E.P. (2004). Image quality assessment: From error measurement to structural similarity. *IEEE Transactions on Image Processing*, 13(1), 1–14.
19. Raja, N.S.M., Rajinikanth, V., & Latha, K. (2014). Otsu based optimal multilevel image thresholding using firefly algorithm. *Modelling and Simulation in Engineering*, 2014, 794574.
20. Satapathy, S.C., Raja, N.S.M., Rajinikanth, V., Ashour, A.S., & Dey, N. (2018). Multilevel image thresholding using Otsu and chaotic bat algorithm. *Neural Computing and Applications*, *29(12)*, 1285–1307.

6 Assessment of Images with Constraints

6.1 SELECTION OF TEST IMAGES

To demonstrate the optimal threshold selection process, it is necessary to choose a test image associated with various constraints, such as complex texture pattern, maximal pixel distribution, nonlinear histogram with complex peak-valley, and associated abnormalities. In the literature, a considerable number of test images exist, and in this work, the Mandrill test image with dimensions $512 \times 512 \times 1$ and $512 \times 512 \times 3$ are considered for initial experimental investigation and then a few more test images available in the literature are also considered for demonstration [1-8].

6.2 ABNORMALITIES IN TEST IMAGES

In digital images, the major abnormality arises due to the association of noise with image pixels. Normally, noise in images can be accounted as the abnormal information available on or next to the pixel of our choice [9–12]. This information will create unwanted troubles during a considerable image processing operation. It will also degrade the performance when an automated analysis or detection is implemented. Hence, noise removal is always preferred to treat and convert the abnormal image into a normal image of our choice. Further, in order to demonstrate the performance of chosen thresholding operation and to evaluate its robustness, it is essential to consider images stained with a measurable noise. A measurable noise has predefined amplitude, mean, and variance [13–16]. The MATLAB software provides information of various noises along with their description. In the proposed work, following noises are considered for evaluation:

- Gaussian: It is a white noise with a predefined mean and variance. Normally, the mean value is assumed as zero and the variance can be measured using a chosen procedure. The variance will be in the order of 0.01.
- Local variance: It is also a type of white noise with a fixed mean and variance. In this, the array dimension of localvariance is equal to the array dimension of the image.
- Poisson: This noise is a generated noise as per the information in image data. The mean and variance of this noise depend on the pixel value of the image.
- Salt & Pepper: This noise has fixed noise density. It can be added intentionally to an image using a chosen density value.
- Speckle: It is a multiplicative noise in an image with zero mean and chosen variance of value 0.04.

6.3   TEST IMAGE STAINED WITH NOISE

This section presents the impact of noise on the image quality and change of its histogram pattern;an experimental demonstration is presented using the images stained with artificial noises.

6.3.1   GAUSSIAN NOISE

The artificial GN with mean 0 and variance 0.01 is introduced on the chosen test image (Mandrill), and attained results are depicted in Figure 6.1. Figure 6.1 (a) presents the original and noise stained images and Figure 6.1 (b) shows the histogram. When the GN is introduced, the image quality considerably degrades and the pixel distribution reduces. From the gray histogram, it can be noted that the maximum pixel level (peak) of the original test image is up to 800, and in the noise-stained case, its value is <600. Further, the pixel distribution in noise-stained image is uniform and its threshold coverage is from Th = 0 to Th = 255. This will increase the complexity during optimal threshold selection process. Similar results are attained for RGB class images. This result confirms that the added noise will degrade the image quality and improve the complexity in histogram-assisted thresholding. Similar results are attained for other artificial noise cases considered in this study.

6.3.2   LOCAL VARIANCE NOISE

To demonstrate the impact of LVN on the test image, the LVN with a white noise of 0 mean and a fixed local variance are added to the image, and the impact of this noise is presented in Figure 6.2. The impact of LVN on the gray/RGB scale image is high as compared to GN. Hence, the complexity in the image is comparatively high. The histogram distribution of this case is also similar to the GN.

6.3.3   POISSON NOISE

The impact of PN on the Mandrill image is then studied by introducing a generated noise based on the image pixel value. The impact of PN on the image and histogram is quite less compared to GN and LVN, and the noise-stained images with PN are depicted in Figure 6.3.

6.3.4   SALT & PEPPER NOISE

The SPN with a fixed density of 0.05 is then added to the image to examine the impact of SPN on gray-/RGB-scaled images. This noise will reduce the visibility of the image by adding a considerable number of dark and white pixels. This noise degrades the image visibility and the impact of SPN on the histogram is very less. This noise will produce a false result when a texture-based image evaluation is performed. This noise can be easily filtered by employing a chosen image filter. The result of SPN in the image can be found in Figure 6.4.

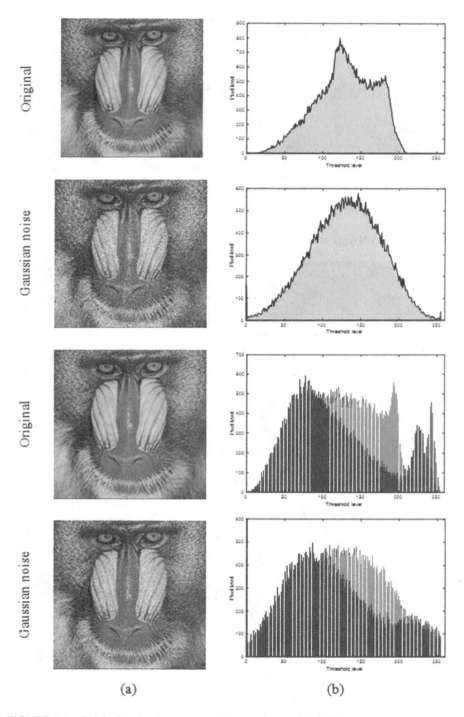

FIGURE 6.1 Original and noise corrupted image of grayscale/RGB cases.

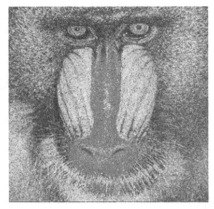

FIGURE 6.2 LVN noise stained imange of Mandrill.

FIGURE 6.3 PN noise stained image of Mandrill.

FIGURE 6.4 SPN noise stained image of Mandrill.

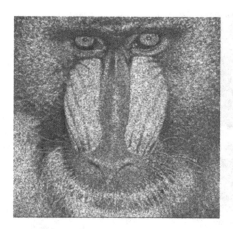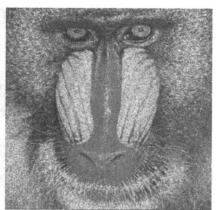

FIGURE 6.5 SN-noise-stained image of Mandrill.

6.3.5 SPECKLE NOISE

SN is a multiplicative noise, and in this study, SN with a 0 mean and a variance of 0.04 is added on the test image and the results are presented in Figure 6.5. Compared to the GN and SPN, the SN considerably degrades the histogram as well as the visibility of the test image. The identification of the optimal threshold with a chosen computerized algorithm may need more computation time to analyze the image corrupted with SN compared to other noises considered in this chapter.

6.4 IMPACT OF NOISE IN THRESHOLDING OUTCOME

A simulated study is performed and results are presented to demonstrate the impact of noise on benchmark test images of dimension $512 \times 512 \times 1$. For this study, well-known benchmark images such as Barbara, Goldhill, House, Butterfly, and Mandrill are considered, and the quality of these images is degraded using the GN and SPN. Later, the original and noise-stained images are evaluated using Firefly Algorithm [1] and Otsu function based multilevel thresholding process discussed in [17–20]. During this study, the following algorithm parameters are assigned for the FA; number of agents = 25, OF = maximization of Otsu's between-class variance, dimension of search = number of thresholds (i.e.,Th = 2,3,4,5), maximum iteration = 3000, and stopping criteria = maximal OF.

Figure 6.6 depicts the original and noise-stained images considered in this study. The corresponding threshold distribution is depicted in Figure 6.7. From Figure 6.7 (a)–(c), it can be seen that the impact of the GN in the histogram is high compared to the impact of SPN. The GN considerably has degraded the pixel density of the test image, and examination of the GN-stained picture is quite difficult compared to the SPN-stained picture. The impact of the noise in PSNR (dB) is computed with a comparative study between the original and noisy image and its values are presented in Table 6.1. From this, it is clear that for each image, the PSNR of the GN is more compared to SPN. This measure can be considered as the noise index.

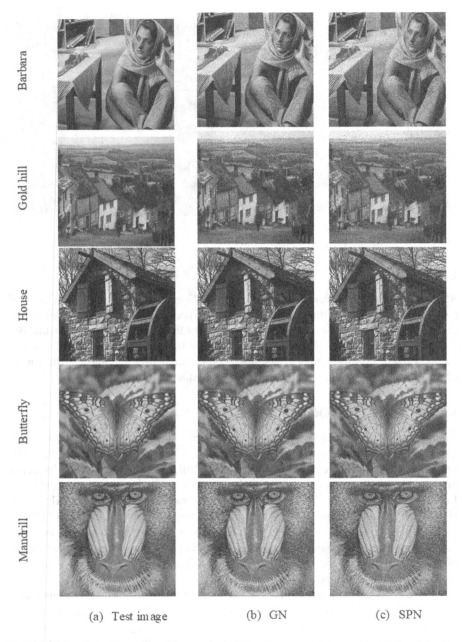

FIGURE 6.6   Experimental results attained with a chosen test image of size $512 \times 512 \times 1$.

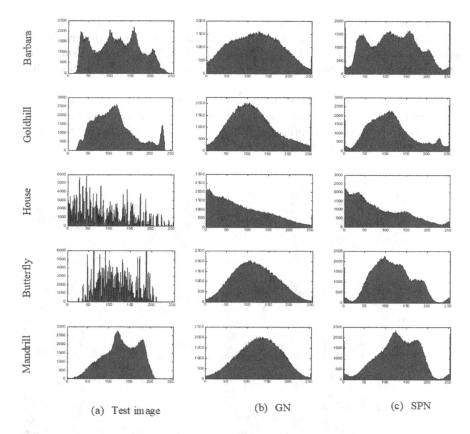

FIGURE 6.7 Histogram of original and noise stained images.

TABLE 6.1
PSNR Value of Noise-Stained Image

Image	Gaussian	SAP
Barbara	19.6045	18.1456
Gold hill	19.5870	18.2714
House	20.2186	17.5030
Butterfly	19.4787	18.4407
Mandrill	19.4670	18.4341

The multithresholding with Th = 2, 3, 4, and 5 is then executed on these images, and the corresponding qualitative results are presented in Figures 6.8 and 6.9. The qualitative results are represented in Tables 6.2 and 6.3. From these results, it can be confirmed that the proposed FA+Otsu-based multithresholding works well on noise-stained images.

The experimental outcome of this study confirms that HA-based methods are robust and offers better result on the original and noise-stained image cases. In the future, the robustness of the proposed procedure can be confirmed by considering the RGB and gray images stained with PN, SN, and LVN.

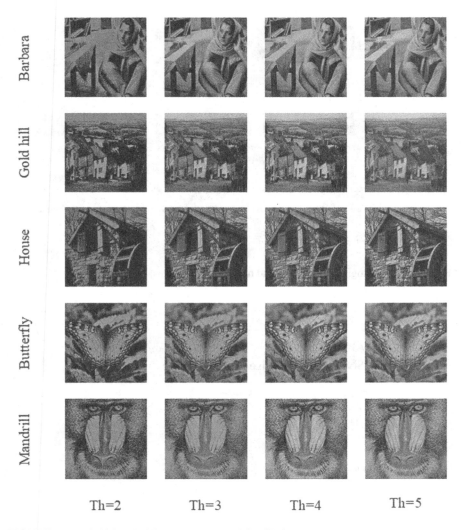

FIGURE 6.8 Multithresholding result attained for GN images.

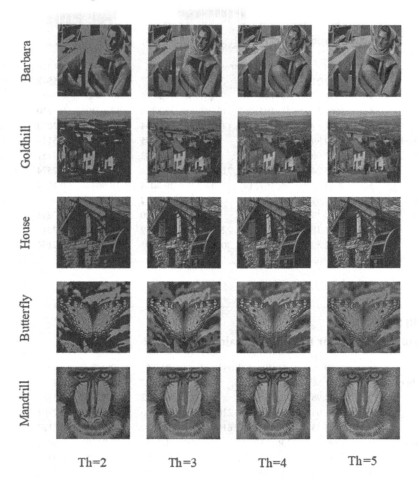

Th=2   Th=3   Th=4   Th=5

FIGURE 6.9   Multithresholding result attained for SAP images.

TABLE 6.2
Results Attained for the Images Stained with GN

	Th	OF	Optimal Threshold	RMSE	PSNR	NAE	NCC
Barbara	2	2513.65	115, 186	66.1561	11.7194	0.4843	0.5723
	3	2663.01	92, 167, 214	47.2121	14.6497	0.3429	0.7232
	4	3069.09	64, 115, 163, 230	34.3875	17.4028	0.2453	0.7908
	5	3110.49	55, 89, 131,191, 241	28.7095	18.9703	0.2063	0.8250
Gold hill	2	2002.19	122, 179	71.4662	11.0488	0.5553	0.5267
	3	2493.79	93, 175, 202	50.1274	14.1293	0.3845	0.6895
	4	2493.79	84, 140, 178, 217	40.8911	15.8982	0.3053	0.7649
	5	2661.36	47, 97, 139, 202, 243	28.5555	19.0170	0.2180	0.8169

(*Continued*)

TABLE 6.2 (Continued)

	Th	OF	Optimal Threshold	RMSE	PSNR	NAE	NCC
House	2	3042.21	98, 172	55.7183	13.2108	0.5542	0.5138
	3	3571.05	56, 141, 185	39.8144	16.1300	0.3921	0.6678
	4	3664.35	41, 103, 153, 216	32.8358	17.8039	0.3110	0.7240
	5	3671.59	46, 101, 139, 171, 239	26.3654	19.7101	0.2638	0.8038
Butterfly	2	1679.89	123, 186	69.2775	11.3190	0.5174	0.5602
	3	1912.98	83, 139, 202	43.6502	15.3311	0.3157	0.7276
	4	1848.24	77, 125, 181, 215	35.6424	17.0915	0.2574	0.7946
	5	2264.32	56, 101, 14, 175, 240	26.5074	19.6635	0.1929	0.8398
Mandrill	2	1104.49	126, 201	69.4477	11.2976	0.4624	0.5998
	3	1694.39	83, 144, 206	43.1620	15.4288	0.2841	0.7462
	4	2121.18	88, 136, 179, 224	36.6459	16.8503	0.2317	0.8177
	5	2072.82	85, 121, 139, 200, 242	34.7405	17.3141	0.2146	0.8282

TABLE 6.3
Results Attained for the Images Stained With SPN

	Th	OF	Optimal Threshold	RMSE	PSNR	NAE	NCC
Barbara	2	1857.56	120, 194	67.5302	11.5408	0.4924	0.5717
	3	1908.98	89, 164, 205	46.2712	14.8246	0.3356	0.7238
	4	3208.04	73, 131, 184, 216	36.6446	16.8506	0.2701	0.7834
	5	3395.33	62, 93, 138, 181, 236	29.7721	18.6546	0.2124	0.8295
Gold hill	2	1996.17	136, 186	81.4050	9.9178	0.6436	0.4571
	3	2458.75	93, 165, 222	50.3458	14.0915	0.3829	0.6863
	4	2685.68	71, 136, 189, 228	38.8902	16.3340	0.2980	0.7594
	5	2882.76	65, 112, 150, 185, 240	31.6276	18.1295	0.2341	0.8110
House	2	2107.81	100, 166	57.5671	12.9273	0.5579	0.5062
	3	3247.69	80, 172, 202	43.1943	15.4223	0.4359	0.6636
	4	3969.19	56, 103, 176, 228	32.6700	17.8478	0.3261	0.7449
	5	3839.55	30, 86, 132, 201, 238	27.0598	19.4843	0.2666	0.7920
Butterfly	2	1664.98	124, 174	72.2234	10.9572	0.5375	0.5423
	3	1706.37	90, 154, 192	47.5288	14.5917	0.3416	0.7154
	4	2058.19	58, 118, 163, 204	34.4348	17.3908	0.2442	0.7859
	5	2213.87	57, 96, 139, 231, 242	29.7852	18.6508	0.2125	0.8063
Mandrill	2	2411.71	125, 206	69.4437	11.2982	0.4574	0.6005
	3	1972.66	93, 153, 232	44.0350	15.2548	0.2828	0.7568
	4	2036.63	88, 131, 168, 24	36.1079	16.9787	0.2207	0.8210
	5	1530.91	48, 94, 144, 179, 250	28.0658	19.1673	0.1807	0.8408

6.5 SUMMARY

This chapter presented the information regarding various constraints existing in test images and their evaluation procedure. This chapter also presented details regarding various noises existing in the literature to test the robustness of the thresholding process and the impact of the noise on gray/RGB images. The experimental demonstration confirmed that the Otsu + FA-based multi thresholding works well on a class of test images attained with the GN and SPN. This study also confirms that the impact of the noise can be found in the image texture as well as the histogram of the image.

REFERENCES

1. Raja, N.S.M., Rajinikanth, V., & Latha, K. (2014). Otsu based optimal multilevel image thresholding using firefly algorithm. *Modelling and Simulation in Engineering, 2014* 794574.
2. Paul, S., & Bandyopadhyay, B. (2014). A Novel Approach for Image Compression Based on MultiLevel Image Thresholding Using Shannon Entropy and Differential Evolution, In: *IEEE Students' Technology Symposium,* 56–61, IEEE Press, USA.
3. Kapur, J.N., Sahoo, P.K., & Wong, A.K.C. (1985). A new method for gray-level picture thresholding using the entropy of the histogram. *Computer Vision Graphics and Image Processing, 29,* 273–285.
4. Manic, K.S., Priya, R.K., & Rajinikanth, V. (2016). Image multithresholding based on Kapur/Tsallis entropy and firefly algorithm. *Indian Journal of Science and Technology, 9*(12), 89949.
5. Agrawal, S., Panda, R., Bhuyan, S., & Panigrahi, B.K. (2013). Tsallisentropy based optimal multilevel thresholding using cuckoo search algorithm. *Swarm and Evolutionary Computation, 11,* 16–30.
6. Ghamisi, P., Couceiro, M.S., & Benediktsson, J.A. (2013). Classification of hyperspectral images with binary fractional order Darwinian PSO and random forests. *SPIE Remote Sensing, 88920,* 88920S.
7. Lee, S.U., Chung, S.Y., & Park, R.H. (1990). A comparative performance study techniques for segmentation. *Computer Vision Graphics Image Processing, 52*(2), 171–190.
8. Sezgin, M., & Sankar, B. (2004). Survey over image thresholding techniques and quantitative performance evaluation. *Journal of Electron Imaging, 13*(1), 146–165.
9. Abhinaya, B., & Raja, N.S.M. (2015). Solving multilevel image thresholding problem—Ananalysis with cuckoo search algorithm, information systems design and intelligent applications. *Advances in Intelligent Systems and Computing, 339,* 177–186.
10. Akay, B. (2013). A study on particle swarm optimization and artificial bee colony algorithms for multilevel thresholding. *Applied Soft Computing Journal, 13*(6), 3066–3091.
11. Rajinikanth, V., & Couceiro, M.S. (2015). RGB histogram based color image segmentation using firefly algorithm. *Procedia Computer Science, 46,* 1449–1457.
12. Manikantan, K., Arun, B.V., & Yaradonic, D.K.S. (2012). Optimal multilevel thresholds based on tsallis entropy method using Golden ratio particles warm optimization for improved image segmentation. *Procedia Engineering, 30,* 364–371.
13. Bhandary, A., Prabhu, G.A., Rajinikanth, V., Thanaraj, K.P., Satapathy, S.C., Robbins, D.E., Shasky, C., Zhang, Y.D., Tavares, J.M.R.S., & Raja, N.S.M. (2020). Deep-learning framework to detect lung abnormality–A study with chest X-Ray and lung CT scan images. *Pattern Recognition Letters, 129,* 271–278.
14. Stark, J.A. (2000). Adaptive image contrast enhancement using generalizations of histogram equalization. *IEEE Transactions on Image Processing, 9*(5), 889–896.

15. Ghamisi, P., Couceiro, M.S., Martins, F.M.L., & Benediktsson, J.A. (2014). Multilevel image segmentation based on fractional-order Darwinian particle swarm optimization. *IEEE Transaction on Geoscience and Remote Sensing, 52(5)*, 2382–2394.

16. Satapathy, S.C., Raja, N.S.M., Rajinikanth, V., Ashour, A.S., & Dey, N. (2018). Multilevel image thresholding using Otsu and chaotic bat algorithm. *Neural Computing and Applications, 29(12),* 1285–1307.

17. Rajinikanth, V., Satapathy, S.C., Fernandes, S.L., & Nachiappan, S. (2016). Entropy based segmentation of tumor from brain MR Images–A study with teaching learning based optimization. *Pattern Recognition Letters, 94*, 87–94.

18. Oliva, D., Cuevas, E., Pajares, G., Zaldivar, D., & Perez-Cisneros, M. (2013). Multilevel thresholding segmentation based on harmony search optimization. *Journal of Applied Mathematics, 2013*, 575414.

19. Ghamisi, P., Couceiro, M.S., Martins, F.M.L., & Benediktsson, J.A. (2014). Multilevel image segmentation based on fractional-order Darwinian particle swarm optimization. *IEEE Tranactionon Geoscience and Remote Sensing, 52(5)*, 2382–2394.

20. Sathya, P.D., & Kayalvizhi, R. (2011). Modified bacterial foraging algorithm based multilevel thresholding for image segmentation. *Engineering Applications of Artificial Intelligence, 24*, 595–615.

7 Thresholding of Benchmark Images

7.1 SELECTION OF TEST IMAGE

The conventional thresholding procedure implemented in a class of application can be found in Figure 7.1. The objective of this technique is to enhance the visibility of the test image based on a suitable pixel grouping concept. The eminence of image outcome depends mainly on the chosen OF and HA. The OF ensures the quality of thresholded image and the HA helps to reduce the computation burden. The technique shown in the figure is common for both gray and RGB scaled test images of varied dimensions and pixel distributions [1–10].

The choice of an appropriate test image to test the proposed technique is essential. Earlier works in the literature can help to choose test images with nonlinear and complex histogram patters. Test images of varied dimensions are attained form well-known image databases. In this work, test images with dimensions $481 \times 321 \times 1$ (Butterfly, Starfish, Snake, Zebra), $321 \times 481 \times 1$ (Bird, Boat, Bridge, Snake), and $512 \times 512 \times 1$ (Hunter, Jet, House, Goldhill) are considered for the examination [11–16].

7.2 SELECTION OF APPROPRIATE THRESHOLDING METHOD

In the image thresholding literature, a considerable number of image threshold techniques are proposed to process a variety of real time images. The choice of an appropriate threshold technique needs some expertise regarding the outcome of process [17–20].

Earlier works in the literature confirm that the Otsu and Kapur are commonly implemented image threshold techniques to process test images. Earlier works also confirm that the attainment of IQC is the prime choice, then we can choose Otsu's function and the enhancement of the image abnormality is a major concern, then we can implement Kapur's technique. In this chapter, the performance of Otsu's and Kapur's technique is verified using a class of test images for Th = 2, 3, 4, and 5. A comparison between the original and threshold images is also performed to confirm the eminence of the implemented threshold operation [21, 22].

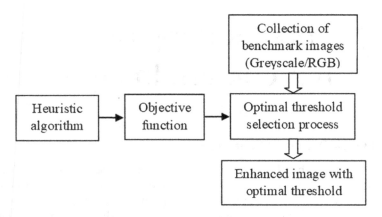

FIGURE 7.1   Overview of the histogram-based thresholding.

7.3   SELECTION OF OBJECTIVE FUNCTION

The selection of a suitable OF is essential to get a convincing result from the auto-mated threshold selection procedure. As discussed earlier, existing OFs are classified as (i) single and (ii) multiple OF.

The choice of OF depends on the computation time to be allotted, throughput needed, and quality requirement. For most of image cases, the threshold imple-mented with a single OF provides better result with lesser experimental time. If better IQV is needed, then a multiple OF controlled by a weighting parameter can be employed. In this chapter, Otsu and Kapur functions are considered for the experi-mental investigation, and due to the simplicity, the single OF presented in Eqns. 7.1 and 7.2 are considered in this work

$$\text{Otsu}_{\max} = J(Th) = \vartheta_0 + \vartheta_1 + \cdots + \vartheta_{L-1} \tag{7.1}$$

$$\text{Kapur}_{\max}(Th) = \sum_{p=1}^{L-1} H_j^C. \tag{7.2}$$

After selecting OF values, the proposed work is executed with CS with the fol-lowing parameters: number of nest = 30, search dimension = assigned threshold (in this work the required thresholds are chosen as Th = 2, 3, 4, 5 and based on these values, the CS search dimension also vary from 2 to 5), number of iterations = 2000 and the stopping criterion = attainment of OF or the maximal iteration number.

7.4 ASSESSMENT OF OUTCOME AND COMPARISON

The executed operation will provide the thresholded images for a chosen threshold. After obtaining the essential outcome from the implemented process, the eminence of the outcome is to be computed based on the achieved IQV. In this work, the common IQVs, such as RMSE, PSNR (dB), NCC, AD, SC, NAE, and SSIM, are computed by executing a pixel-wise comparison between the original image and thresholded image. Based on these values, the performance of the Otsu as well as Kapur functions is confirmed.

The remaining part of this section presents results attained with the proposed technique for various test image cases. This work clearly demonstrates outcomes, such as the maximized OF, optimal thresholds identified, and computed IQV for Otsu as well as Kapur functions.

Figure 7.2 (a) presents the grayscale benchmark image of dimension $481 \times 321 \times 1$ along with its gray threshold (Figure 7.2 (b)). Later, Otsu's- and Kapur's-based thresholding is executed on these images for various thresholds, and the attained results are recorded. Figure 7.3 depicts the optimal results attained for every image with various threshold values, and the corresponding quality parameters measures from these images are depicted in Table 7.1 (for Otsu) and Table 7.2 (for Kapur). After getting the essential results, it is necessary to analyze the values to confirm that the increase in threshold value improves the IQV. Figure 7.4 presents the graphical comparison of the attained value for the butterfly image and this result confirms that the increase in threshold improved the quality of thresholded image. Figure 7.5 presents a Glyph-plot for all existing measures in Table 7.2 in which the subplots 1–4, 5–8, 9–12, and 13–16 present results of Butterfly, Starfish, Snake, and Zebra, respectively for various chosen threshold values.

A similar procedure is then performed using the test images of size $321 \times 481 \times 1$ (Figure 7.6), and the corresponding outcomes are presented in Figure 7.7, Table 7.3, and Table 7.4. From these results, it can be noted that the CS-based optimal threshold selection works well with the single OF and results attained with Otsu's and Kapur's functions are approximately similar.

Later, the considered procedure is executed on the benchmark image of size $512 \times 512 \times 1$ (Figure 7.8). The attained results are presented in Figure 7.9 and Tables 7.5 and 7.6. Figures 7.10 and 7.11 present the box-plot comparison of table values 7.5 and 7.6, respectively. This plot confirms that the result attained on the chosen test image with Otsu and Kapur are approximately similar. If this procedure is executed with a multiple OF, then we can enhance the IQV attained from the thresholded image.

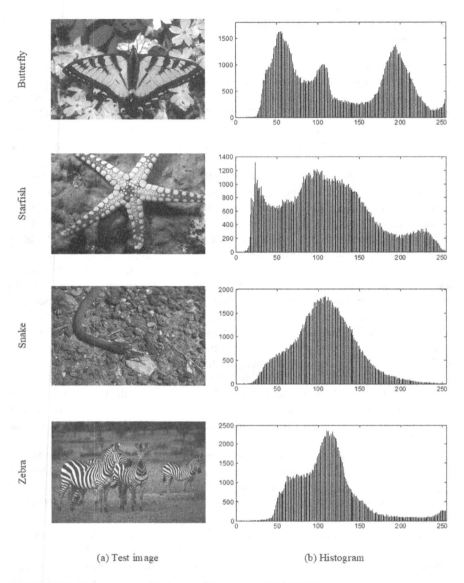

(a) Test image   (b) Histogram

FIGURE 7.2   Benchmark test images of dimension $481 \times 321 \times 1$.

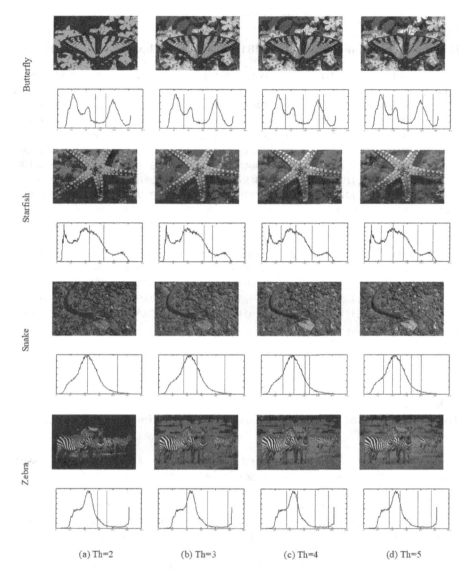

(a) Th=2   (b) Th=3   (c) Th=4   (d) Th=5

FIGURE 7.3 Thresholding result attained for the benchmark test images of dimension $481 \times 321 \times 1$ with Otsu's function.

TABLE 7.1
Threshold Outcome Obtained for 481× 321× 1 Sized Images with Otsu

Image	CF	OT	RMSE	PSNR	NCC	AD	SC	NAE	SSIM
Butterfly	3515.67	134, 170	63.6487	12.0550	0.6886	54.9351	1.7126	0.4426	0.3494
	3871.61	84, 155, 199	42.4670	15.5698	0.8025	36.5798	1.4347	0.2947	0.5101
	3986.31	81, 144, 199, 215	42.0776	15.6498	0.8041	36.2108	1.4317	0.2917	0.5150
	4044.06	64, 97, 150, 202, 239	33.7228	17.5723	0.8369	28.2169	1.3663	0.2273	0.5951
Starfish	1984.03	115, 165	60.8285	12.4487	0.6414	51.1049	1.8833	0.4687	0.3094
	2545.10	85, 157, 187	43.3644	15.3881	0.7373	37.1024	1.6644	0.3403	0.4395
	2779.18	68, 119, 177, 234	33.5159	17.6258	0.7929	29.0624	1.5126	0.2665	0.5417
	2862.60	60, 101, 138, 187, 241	27.7335	19.2707	0.8323	23.7597	1.3963	0.2179	0.6157
Snake	856.80	109, 215	63.8989	12.0209	0.5736	54.1160	2.1681	0.5023	0.3671
	1117.86	87, 134, 231	41.6637	15.7356	0.7383	33.8661	1.6391	0.3143	0.5887
	1227.52	76, 114, 154, 170	31.5006	18.1644	0.8129	25.1095	1.4242	0.2331	0.7076
	1285.15	69, 101, 129, 166, 201	25.8765	19.8727	0.8511	20.2537	1.3267	0.1880	0.7786
Zebra	948.77	145, 178	96.4799	8.4421	0.2997	89.9205	3.9604	0.8021	0.1697
	1394.16	100, 173, 243	52.6092	13.7096	0.6968	42.8582	1.7014	0.3823	0.4599
	1525.22	91, 129, 190, 237	42.9975	15.4619	0.7695	33.9183	1.4952	0.3026	0.5665
	1579.81	83, 111, 141, 198, 252	35.9651	17.0132	0.8188	26.5957	1.3729	0.2372	0.6594

TABLE 7.2
Threshold Outcome Obtained for 481× 321× 1 Sized Images with Kapur's Function

Image	CF	OT	RMSE	PSNR	NCC	AD	SC	NAE	SSIM
Butterfly	13.1647	123, 256	74.5635	10.6803	0.5290	69.6720	2.9265	0.5613	0.2953
	18.1718	116, 174, 255	59.1605	12.6902	0.7294	50.3295	1.5688	0.4055	0.3626
	22.4942	69, 121, 173, 254	38.7783	16.3590	0.8003	33.8441	1.4765	0.2727	0.5607
	26.6415	79, 135, 174, 225, 255	39.4059	16.2196	0.8209	33.7986	1.3869	0.2723	0.5487
Starfish	13.5349	159, 255	90.1475	9.0317	0.3806	79.7642	3.2652	0.7316	0.1371
	18.7058	89, 172, 255	48.6062	14.3970	0.6827	42.3447	1.9088	0.3884	0.3980
	23.1702	78, 126, 177, 249	36.2815	16.9371	0.7810	30.7746	1.5379	0.2822	0.5134
	27.3599	63, 98, 137, 194, 254	28.4630	19.0452	0.8273	24.1876	1.4107	0.2218	0.6109
Snake	12.8439	169, 256	105.0236	7.7051	0.1362	98.4500	8.2811	0.9138	0.0510
	17.8971	87, 169, 255	47.0211	14.6849	0.6701	40.3110	1.9599	0.3742	0.5034
	22.4163	80, 148, 192, 251	41.2903	15.8138	0.7174	35.3819	1.7670	0.3284	0.5820
	26.5775	73, 115, 151, 201, 249	30.6488	18.4025	0.8123	25.0318	1.4351	0.2323	0.7217
Zebra	13.0677	163, 252	102.7331	7.8966	0.2253	97.2504	5.2316	0.8675	0.1242
	17.7987	100, 165, 254	53.3856	13.5823	0.6808	43.6177	1.7804	0.3891	0.4594
	22.1342	88, 138, 186, 246	42.7068	15.5209	0.7550	35.0611	1.5676	0.3128	0.5648
	26.1548	85, 152, 181, 217, 254	43.4223	15.3765	0.7356	36.7876	1.6568	0.3282	0.5436

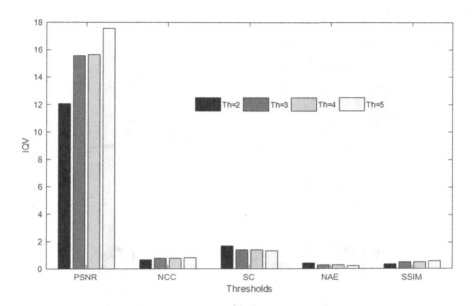

FIGURE 7.4 Comparison of the IQV with respect to various thresholds (Th = 2, 3, 4, 5) for butterfly image.

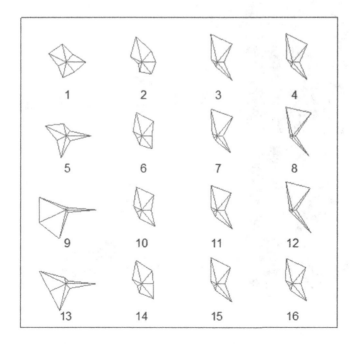

FIGURE 7.5 Comparison of the IQV with respect to various thresholds (Th = 2, 3, 4, 5) using Glyph-plot.

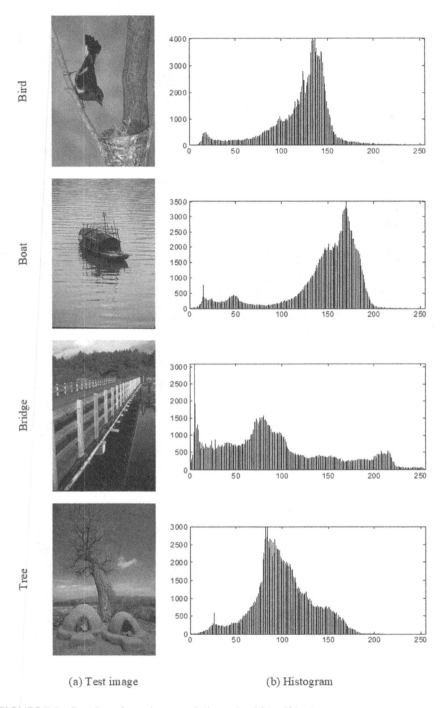

(a) Test image   (b) Histogram

FIGURE 7.6   Benchmark test images of dimension $321 \times 481 \times 1$.

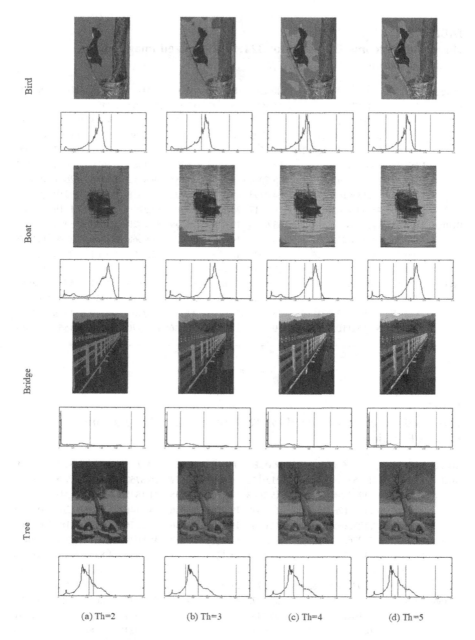

(a) Th=2 (b) Th=3 (c) Th=4 (d) Th=5

FIGURE 7.7 Thresholding result attained for the benchmark test images of dimension $321 \times 481 \times 1$ with Otsu's function.

TABLE 7.3
Threshold Outcome Obtained for 321× 481× 1 Sized Images with Otsu's Function

Image	CF	OT	RMSE	PSNR	NCC	AD	SC	NAE	SSIM
Bird	715.29	101,178	44.9677	15.0728	0.7072	38.6162	1.8406	0.3198	0.6653
	900.44	71,122,221	28.2765	19.1023	0.8217	23.4505	1.4403	0.1942	0.7483
	974.49	64,111,140,241	23.2705	20.7947	0.8618	18.7830	1.3191	0.1556	0.7764
	1026.63	61,104,131,164,218	19.6694	22.2550	0.8854	15.9717	1.2572	0.1323	0.8200
Boat	1357.21	105,207	57.3746	12.9564	0.6390	53.4085	2.3679	0.3685	0.4949
	1614.30	89,154,248	35.5034	17.1254	0.8185	29.8752	1.4445	0.2061	0.7069
	1681.82	78,133,162,192	24.0793	20.4979	0.8922	18.9828	1.2349	0.1310	0.7727
	1709.40	41,90,136,163,237	19.6717	22.2540	0.8999	16.2206	1.2243	0.1119	0.8183
Bridge	2597.03	110,254	63.6203	12.0589	0.4579	53.1420	3.3233	0.6495	0.2098
	3383.56	54,135,255	38.6042	16.3981	0.6629	31.0068	2.1381	0.3790	0.4807
	3563.30	49,108,169,226	30.1020	18.5589	0.7606	25.3301	1.6461	0.3096	0.5356
	3659.42	32,71,116,173,243	23.1447	20.8418	0.8097	18.7901	1.4917	0.2297	0.6376
Tree	602.50	107,122	66.9586	11.6147	0.5319	56.0836	2.1371	0.5577	0.2854
	788.32	74,118,252	31.4041	18.1911	0.7671	25.8428	1.6050	0.2570	0.6706
	873.78	65,99,132,221	23.9379	20.5491	0.8258	19.8900	1.4219	0.1978	0.7512
	908.49	58,88,110,138,242	19.2182	22.4566	0.8646	15.6170	1.3116	0.1553	0.8013

TABLE 7.4
Threshold Outcome Obtained for 321× 481× 1 Sized Images with Kapur's Function

Image	CF	OT	RMSE	PSNR	NCC	AD	SC	NAE	SSIM
Bird	12.3953	171,253	121.9126	6.4098	0.0477	116.8752	23.4927	0.9680	0.0240
	17.9022	91,173,254	45.4683	14.9766	0.6745	41.1848	2.0803	0.3411	0.6718
	22.2719	67,113,176,251	27.8645	19.2298	0.8086	24.6146	1.4998	0.2039	0.7886
	26.6950	77,127,165,212,254	27.6168	19.3073	0.8386	22.2976	1.3776	0.1847	0.7565
Boat	12.7880	117,236	50.9501	13.9879	0.6972	45.9024	1.9678	0.3167	0.5479
	17.4350	63,121,245	43.1941	15.4223	0.7331	39.4829	1.8248	0.2724	0.6166
	21.6548	54,103,136, 215	31.3179	18.2150	0.8148	27.7412	1.4869	0.1914	0.7210
	25.5129	58,102,130,158,218	20.4723	21.9075	0.8950	17.2770	1.2371	0.1192	0.8103
Bridge	13.2613	113,255	63.8272	12.0307	0.4592	53.4047	3.2678	0.6527	0.2068
	18.1775	107,164,254	55.3437	13.2694	0.5987	45.3398	2.0453	0.5542	0.2591
	22.4565	106,153,219,255	54.6084	13.3856	0.6147	44.2924	1.9488	0.5414	0.2558
	26.4750	67,99,142,196,253	27.1817	19.4453	0.8251	21.0758	1.3880	0.2576	0.5171
Tree	12.1614	65,237	48.9381	14.3379	0.5637	41.9703	2.9128	0.4174	0.5820
	17.4754	72,209,255	45.6397	14.9440	0.6081	38.3109	2.4750	0.3810	0.5975
	21.8613	50,131,215,251	43.7872	15.3039	0.6351	39.1772	2.2567	0.3896	0.5682
	25.9354	65,98,132,209,246	23.8094	20.5958	0.8266	19.7523	1.4198	0.1964	0.7522

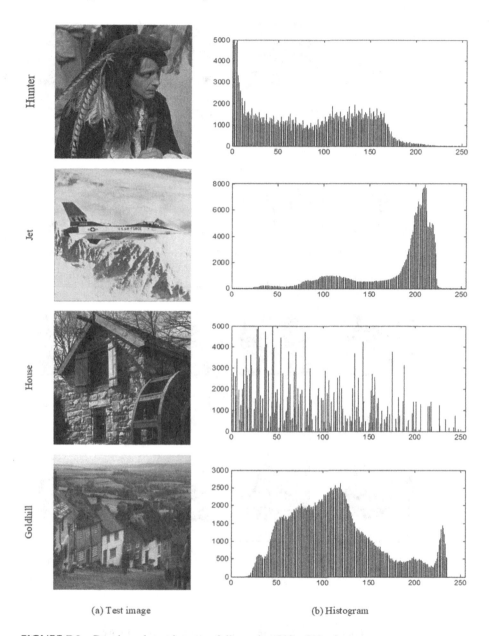

(a) Test image   (b) Histogram

FIGURE 7.8   Benchmark test images of dimension $512 \times 512 \times 1$.

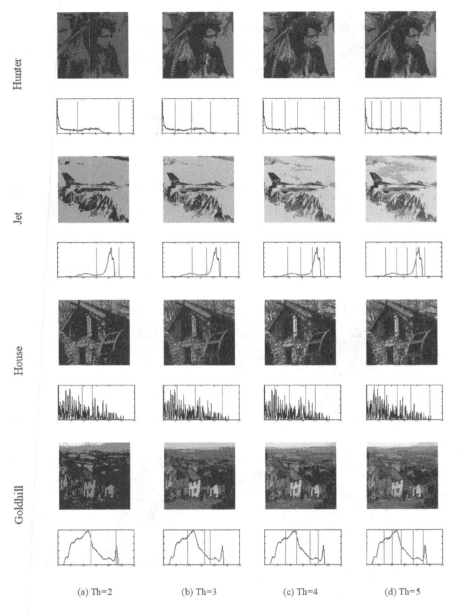

(a) Th=2   (b) Th=3   (c) Th=4   (d) Th=5

FIGURE 7.9 Thresholding result attained for benchmark test images of dimension $512 \times 512 \times 1$ with Otsu's function.

TABLE 7.5
Threshold Outcome Obtained for $512 \times 512 \times 1$ Sized Images with Otsu's Function

Image	CF	OT	RMSE	PSNR	NCC	AD	SC	NAE	SSIM
Hunter	2700.05	80,244	47.2373	14.6451	0.5607	38.1553	2.7967	0.4903	0.4356
	3063.63	51,116,190	32.3267	17.9396	0.7173	25.7913	1.8344	0.3314	0.5937
	3212.40	36,86,135,242	22.7175	21.0036	0.8194	18.2098	1.4423	0.2340	0.7125
	3268.81	27,65,104,143,217	18.6742	22.7060	0.8489	15.0365	1.3611	0.1932	0.7784
Jet	1690.88	154,241	68.6799	11.3942	0.6963	61.5146	1.8842	0.3433	0.6706
	1836.72	116,174,228	40.5151	15.9785	0.8675	29.6046	1.2770	0.1652	0.7969
	1910.80	94,145,190,237	25.7085	19.9293	0.9264	17.0835	1.1467	0.0953	0.8373
	1949.47	80,122,165,200,233	22.7261	21.0003	0.9155	17.7135	1.1817	0.0988	0.8456
House	2931.56	94,134	44.1517	15.2318	0.6585	34.6785	1.9401	0.4508	0.4748
	3420.29	55,127,233	36.5652	16.8694	0.6920	29.2628	1.9231	0.3804	0.6088
	3619.31	42,98,163,200	28.1219	19.1499	0.7771	22.7713	1.5758	0.2960	0.7163
	3724.50	32,74,123,177,252	23.8711	20.5734	0.8053	19.3174	1.4958	0.2511	0.7968
Goldhill	1566.81	131,230	79.4740	10.1263	0.5043	69.9956	2.3290	0.6238	0.2272
	2064.90	94,161,183	48.5066	14.4148	0.7283	40.8784	1.6304	0.3643	0.4956
	2218.11	83,126,179,210	38.8634	16.3400	0.7969	31.7503	1.4401	0.2830	0.6262
	2290.47	69,102,137,185,224	28.7592	18.9553	0.8476	23.0955	1.3326	0.2058	0.7435

TABLE 7.6
Threshold Outcome Obtained for $512 \times 512 \times 1$ Sized Images with Kapur's Function

Image	CF	OT	RMSE	PSNR	NCC	AD	SC	NAE	SSIM
Hunter	12.7876	99,256	46.3087	14.8176	0.6011	36.8594	2.3302	0.4736	0.3983
	17.8209	92,179,253	44.2833	15.2060	0.6170	35.6443	2.2649	0.4580	0.4348
	22.5023	60,117,179,252	30.3233	18.4953	0.7488	24.5272	1.6808	0.3152	0.5910
	26.6467	46,90,130,179,254	22.5560	21.0656	0.8184	18.3784	1.4480	0.2362	0.6928
Jet	12.3673	162,255	67.7400	11.5139	0.7201	58.0823	1.7404	0.3241	0.6643
	17.5645	76,174,243	39.5171	16.1951	0.8171	35.1128	1.4709	0.1959	0.8148
	22.2430	70,128,182,247	28.9825	18.8881	0.8652	26.0481	1.3246	0.1454	0.8700
	26.3680	20,73,134,184,229	27.3678	19.3860	0.8731	24.4178	1.3020	0.1363	0.8781
House	11.5377	88,256	55.9428	13.1759	0.4844	44.8585	3.4817	0.5831	0.3768
	15.5884	57,122,254	38.4537	16.4320	0.6664	30.6954	2.0695	0.3990	0.5895
	19.2068	47,88,149,251	30.2041	18.5295	0.7518	24.3106	1.6764	0.3160	0.6762
	22.4207	47,84,125,168,250	25.1050	20.1356	0.8108	20.4661	1.4584	0.2660	0.7050
Goldhill	13.1919	140,254	86.2620	9.4144	0.4097	78.4238	3.1745	0.6989	0.1862
	18.0935	90,157,253	49.3988	14.2565	0.6865	42.7578	1.8676	0.3811	0.4997
	22.4434	75,136,185,245	39.6218	16.1721	0.7563	34.3810	1.6204	0.3064	0.6177
	26.4929	69,116,150,192,238	32.1518	17.9867	0.8180	26.6301	1.4185	0.2373	0.7086

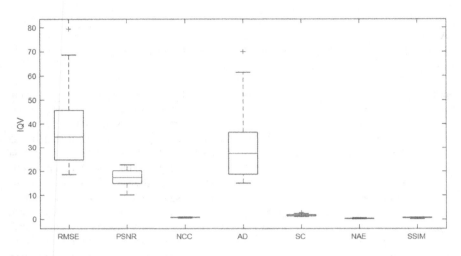

FIGURE 7.10 Box-plot to represent the variation in the IQV of Table 7.5.

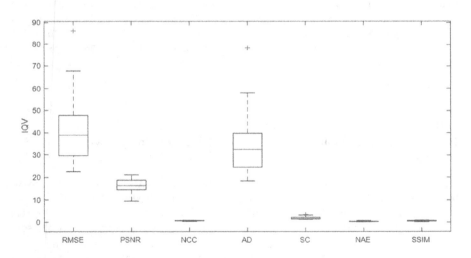

FIGURE 7.11 Box-plot to represent the variation in the IQV of Table 7.6.

7.5 SUMMARY

The initial part of this chapter presents the outline of HA-based thresholding process and also provides the information on existing benchmark test images in the literature. The choice of OF and its implementation is also discussed. Finally, the HA-based threshold operation is demonstrated using the appropriate experimental results obtained using the Otsu and Kapur functions. The use of the Bar-chart, Glyph-plot, and Box-plot is also demonstrated, and the experimental outcome confirms that the results attained for Otsu and Kapur functions with single OF are approximately similar.

REFERENCES

1. Lee, S.U., Chung, S.Y., & Park, R.H. (1990). A comparative performance study techniques for segmentation. *Computer Vision, Graphics, and Image Processing, 52*(2), 171–190.
2. Sezgin, M., & Sankar, B. (2004). Survey over image thresholding techniques and quantitative performance evaluation. *Journal of Electron Imaging, 13*(1), 146–165.
3. Abhinaya, B., & Raja, N.S.M. (2015). Solving multilevel image thresholding problem—An analysis with cuckoo search algorithm, information systems design and intelligent applications. *Advances in Intelligent Systems and Computing, 339*, 177–186.
4. Agrawal, S., Panda, R., Bhuyan, S., & Panigrahi, B.K. (2013). Tsallis entropy based optimal multilevel thresholding using cuckoo search algorithm. *Swarm and Evolutionary Computation, 11*, 16–30.
5. Ghamisi, P., Couceiro, M.S., Martins, F.M.L., & Benediktsson, J.A. (2014). Multilevel image segmentation based on fractional-order Darwinian particle swarm optimization. *IEEE Transaction on Geoscience and Remote Sensing, 52*(5), 2382–2394.
6. Sathya, P.D., & Kayalvizhi, R. (2011). Modified bacterial foraging algorithm based multilevel thresholding for image segmentation. *Engineering Applications of Artificial Intelligence, 24*, 595–615.
7. Sarkar, S., Paul, S., Burman, R., Das, S., & Chaudhuri, S.S. (2014) A fuzzy entropy based multilevel image thresholding using differential evolution. *Lecture Notes in Computer Science, 8947*, 386–395.
8. Anusuya, V., & Latha, P. (2014). A novel nature inspired fuzzy Tsallis entropy segmentation of magnetic resonance images. *Neuroquantology, 12*(2), 221–229.
9. Lakshmi, V.S., Tebby, S.G., Shriranjani, D., & Rajinikanth, V. (2016). Chaotic cuckoo search and Kapur/Tsallis approach in segmentation of T. Cruzi From blood smear images. *International Journal of Computer Science and Information Security, 14*, 51–56.
10. Manikantan, K., Arun, B.V., & Yaradonic, D.K.S. (2012). Optimal multilevel thresholds based on Tsallis entropy method using Golden ratio particle swarm optimization for improved image segmentation, *Procedia Engineering, 30*, 364–371.
11. Available online: http://decsai.ugr.es/cvg/CG/base.htm. (Accessed on 16th August 2020)
12. Available online: http://www.eecs.berkeley.edu/Research/Projects/CS/vision/bsds/BSDS300/html/dataset/images.html. (Accessed on 16th August 2020)
13. Ghamisi, P., Couceiro, M.S., Martins, F.M.L., & Benediktsson, J.A. (2014). Multilevel image segmentation based on fractional-order Darwinian particle swarm optimization. *IEEE Tranaction on Geoscience and Remote Sensing, 52*(5), 2382–2394.
14. Satapathy, S.C., Raja, N.S.M., Rajinikanth, V., Ashour, A.S., & Dey, N. (2018). Multilevel image thresholding using Otsu and chaotic bat algorithm. *Neural Computing and Applications, 29(12)*, 1285–1307.
15. Rajinikanth, V., Satapathy, S.C., Fernandes, S.L., & Nachiappan, S. (2016). Entropy based segmentation of tumor from brain MR Images–A study with teaching learning-based optimization. *Pattern Recognition Letters, 94*, 87–94.
16. Oliva, D., Cuevas, E., Pajares, G., Zaldivar, D., & Perez-Cisneros, M. (2013). Multilevel thresholding segmentation based on harmony search optimization. *Journal of Applied Mathematics, 2013*(575414), 24.
17. Raja, N.S.M., Rajinikanth, V., & Latha, K. (2014). Otsu based optimal multilevel image thresholding using firefly algorithm. *Modelling and Simulation in Engineering, 2014* (794574), 17.
18. Kapur, J.N., Sahoo, P.K., & Wong, A.K.C. (1985). A new method for gray-level picture thresholding using the entropy of the histogram. *Comput Vis Graph Image Process, 29*, 273–285.

19. Manic, K.S., Priya, R.K., & Rajinikanth, V. (2016). Image multithresholding based on Kapur/Tsallis entropy and firefly algorithm. *Indian Journal of Science and Technology*, 9(12), 89949.

20. Akay, B. (2013). A study on particle swarm optimization and artificial bee colony algorithms for multilevel thresholding. *Applied Soft Computing Journal*, 13(6), 3066–3091.

21. Rajinikanth, V., & Couceiro, M.S. (2015). RGB histogram based color image segmentation using firefly algorithm, *Procedia Computer Science*, 46, 1449–1457.

22. Stark, J.A. (2000). Adaptive image contrast enhancement using generalizations of histogram equalization. *IEEE Transactions on Image Processing*, 9(5), 889–896.

8 Thresholding of Biomedical Images

8.1 NEED OF THRESHOLDING FOR MEDICAL IMAGES

In most of the disease cases, the disease in organs is usually assessed using the image-assisted automated/semiautomated computer algorithms. In most of the algorithms, the thresholding operation is employed to preprocess the image, and its outcome is then sent for decision-making process or further analysis.

Due to the availability of modern computing facilities, the implementation of computed-based algorithms is very common in disease detection process, and due to this reason, recently, a number of HA-based machine-learning techniques are proposed and implemented to examine a class of abnormalities in two-dimensional (2D) medical images available as gray/RGB scale [1-3].

Further, this chapter presents the results attained on various 2D medical images processed using trilevel threshold process. In order to automate the threshold process, this work employed the Bat Algorithm (BA)-assisted thresholding with a chosen OF (Otsu/Kapur function). The initial tuning of the BA is as follows: number of agents = 25; search dimension = 3 (tri-level threshold); OF = maximization of the Otsu/Kapur function; maximum iteration value = 3000; and stopping criterion = maximized OF or maximal iteration value. During the medical image examination, the prime objective is enhancing the abnormality and it is not necessary to compute the image quality values, such as RMSE, PSNR, SSIM, NCC, SC, AD, and NAE.

8.2 THRESHOLDING FOR MRI

Magnetic Resonance Imaging (MRI) is a widely preferred imaging modality to examine a class of internal body organs. The main advantage of the MRI is that it offers a reconstructed 3D image and it can be examined in its 3D form or 2D form. Further, the MRI supports various modalities, such as Flair, T1, T1C, T2, fMRI, and Diffusion Weight images. Another advantage of the MRI is using the 3D to 2D conversion, we can get various images, views, such as Axial, Coronal and Sagittal, and based on the requirement, any view can be chosen for the assessment [4-6].

In hospitals, the MRI can be used to examine the abnormality in brain, heart, and breast. In the proposed work, the experimental results from the MRI of brain (form the tumor and stroke cases), heart, and breast are presented, and the essential images are attained form the benchmark image datasets.

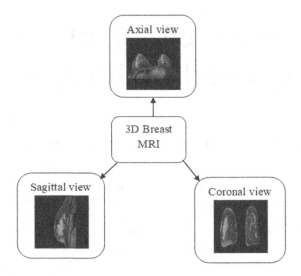

FIGURE 8.1   Conversion of the 3D breast MRI into 2D slices.

- Breast MRI:
  Breast abnormality is one of the leading diseases in women, and breast cancer is one of the acute diseases. The tumor in the breast can be diagnosed using various imaging techniques, such as mammograms, thermal imagery, and MRI. As discussed earlier, MRI provides the 3D view and it can be examined into various 2D slices as depicted in Figure 8.1.

  For most of the medical images, Kapur's technique is the preferred OF and the implementation of a trilevel threshold with Kapur will separate the image into three sections, such as abnormal pixel, background, and normal pixel. Figure 8.2 depicts the test image (a), GT (b), and thresholded image (c). After enhancing the tumor section in the test image, a chosen segmentation procedure is then implemented to extract the abnormal pixel from Figure 8.2 (c). The extracted section is to be compared with Figure 8.2 (b)

(a) Test image   (b) Ground truth   (c) Thresholded image

FIGURE 8.2   Thresholding of chosen 2D MRI slices using Kapur's technique.

to validate the performance of the proposed technique. The image dataset for the breast MRI can be found in references [7, 8].

- Brain MRI:
  The brain is a significant organ in the human physiological system. It responds to process and control the action of other organs of the body according to the processed signals. The abnormality in the brain affects the whole operation. The untreated illness may lead to various problems ranging from mild-disability to death. The early symptoms of brain tumor are invisible and it can be detected only by using a suitable imaging modality, like computed tomography (CT) or MRI. The visibility of the tumor in MRI is less and, hence, it requires a suitable procedure to recognize the size and orientation of the tumor. The earlier works on the brain MRI assessment revealed that the integration of thresholding with a chosen segmentation helps to attain better results. Further, the choice of the thresholding and the segmentation operation needs prior knowledge to attain the improved outcome. In this work, the brain MRI slice obtained from references [9–11] is considered for the analysis, and Kapur's-entropy-based thresholding helped to attain better enhancement of the tumor and the results are depicted in Figure 8.3. In this work, MRI with T2 modality is considered and similar results can be obtained with other MRI slices as depicted in Figure 8.4. The essential images are attained form the BRATS 2015 database [12, 13].

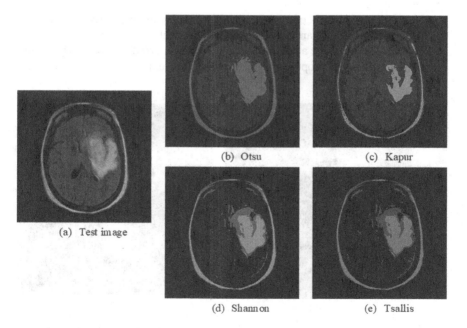

(a) Test image

(b) Otsu

(c) Kapur

(d) Shannon

(e) Tsallis

FIGURE 8.3 Thresholding result obtained for the brain MRI.

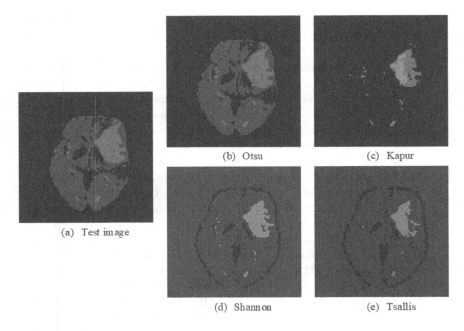

(a) Test image   (b) Otsu   (c) Kapur   (d) Shannon   (e) Tsallis

FIGURE 8.4 Kapur's-entropy-based thresholding to enhance the tumor fragment from other brain section.

- Heart MRI:
  Heart abnormality is one of the common problems in elderly people and the unrecognized heart disease will lead to death. Figure 8.5 depicts the heart MRI test image and the thresholded image. From Figure 8.5 (b) it can be noted that the proposed procedure helps to enhance the heart section

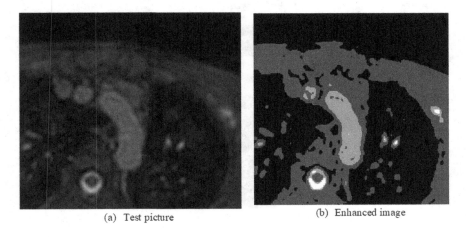

(a) Test picture   (b) Enhanced image

FIGURE 8.5 Otsu based thresholding result attained with the 2D heart MRI slice.

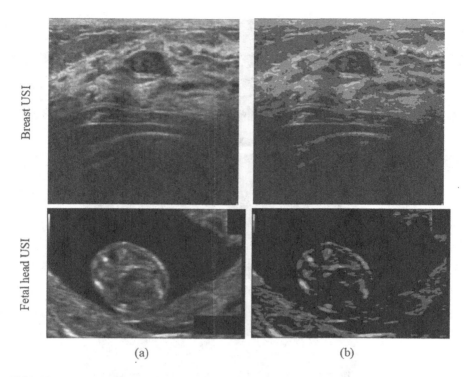

Breast USI

Fetal head USI

(a)   (b)

FIGURE 8.6   Kapur-based trilevel threshold result obtained using the USI.

considerably. The enhanced heart section can be extracted by employing a suitable segmentation procedure discussed in the earlier works. Further, the outcome of this process is to be verified and approved by a doctor. The benchmark heart MRI data considered in this study can be accessed from [14].

8.3   THRESHOLDING FOR ULTRASOUND IMAGE

Ultrasound imaging (USI) is also another commonly used modality to record the internal body organs and their abnormality using the sound wave. After recording the images, the abnormal section is enhanced for further analysis. In this work, Kapur's approach is employed to threshold the image and its outcome is presented in Figure 8.6. The database for further assessment can be found in [15, 16].

8.4   THRESHOLDING FOR RGB-SCALE MEDICAL IMAGES

Like the greyscale image, RGB scale images are widely used in the medical domain to record the disease in body organs. As discussed earlier, the assessment of the RGB scaled images are quite complex. Therefore, a suitable image assessment technique is always essential to separate the ROI from the raw test image. This section presents

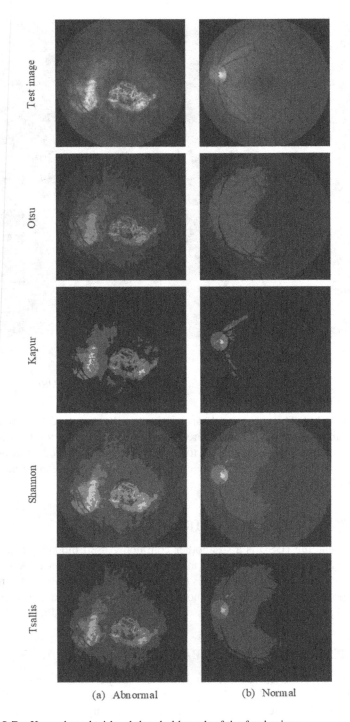

(a) Abnormal   (b) Normal

FIGURE 8.7   Kapur-based tri-level threshold result of the fundus image.

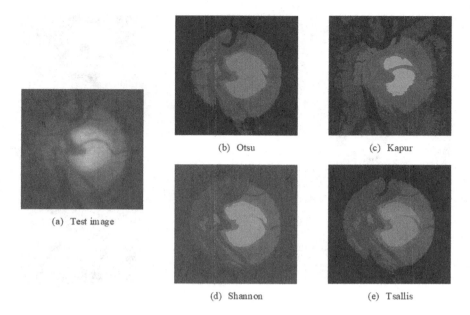

(a) Test image

(b) Otsu

(c) Kapur

(d) Shannon

(e) Tsallis

FIGURE 8.8   Enhancement of retinal optic section.

the assessment of some benchmark medical dataset images and also shares the location of the dataset for future research purpose.

Figure 8.6 presents the experimental outcome attained for a fundus retinal picture of normal/Age-Related Macular Degeneration (AMD) class [17, 18]. The AMD is one of the common diseases in elderly people, and the untreated AMD will affect the vision system badly. The results of Figure 8.6 (b) confirm that Kapur's technique helps to extract the abnormal section with better accuracy.

A similar procedure is executed in the retinal optic assessment task (Figure 8.7) using the RIM-ONE database [19] and the result of the Otsu's technique seems better compared to the results of Kapur's technique.

Kapur-based trilevel thresholding is then executed on other RGB-scaled test images and better results are attained. Figure 8.8 presents the result of dermoscopy images [20, 21], Figures 8.9 and 8.10 present the results attained for the white blood cell database [22] and gastric polyp [23]. From these results, it can be noted that the chosen thresholding procedure with the Otsu's and Kapur's functions works well on a class of medical-grade test images.

8.5   THRESHOLDING OF LUNG CT SCAN SLICE INFECTED WITH COVID-19

The early detection of pneumonia and the probable treatment execution will reduce the death rate in humans. The lung infection due to pneumonia is normally assessed using imaging procedures such as the CT and Chest Radiographs (X-ray). In these procedures, the patient is screened using a chosen imaging method, and after the

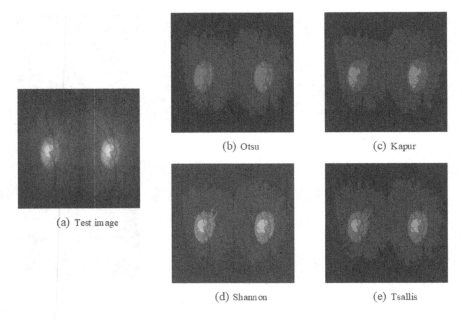

(a) Test image   (b) Otsu   (c) Kapur   (d) Shannon   (e) Tsallis

FIGURE 8.9   Thresholdeddermoscopy images using Kapur's technique.

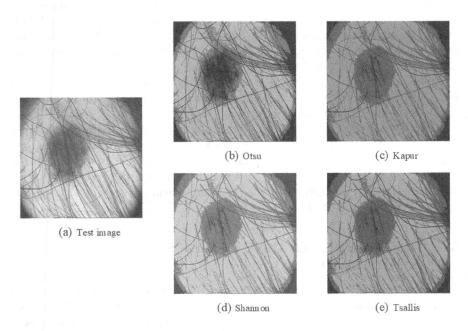

(a) Test image   (b) Otsu   (c) Kapur   (d) Shannon   (e) Tsallis

FIGURE 8.10   White blood cell enhancement with Kapur's thresholding.

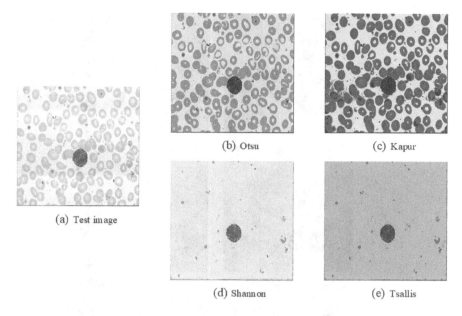

(a) Test image

(b) Otsu

(c) Kapur

(d) Shannon

(e) Tsallis

FIGURE 8.11 Enhancement of gastric polyp with Kapur's technique.

screening process, the doctor will examine the chest images and based on his observation, a possible treatment procedure is implemented. Due to its significance, a number of procedures for detecting pneumonia are already proposed and implemented by the researchers [24–26]. Figure 8.11 and 12 depicts the lung CT scan images and its processing.

Pneumonia due to the Coronavirus Disease (COVID-19) is one of the novel diseases that affected a large human group worldwide. This disease was first discovered in Wuhan, China, in December 2019, and due to its outbreak, the infection rate is gradually increasing. The earlier research works on COVID-19 confirms that

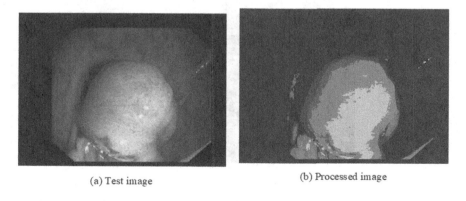

(a) Test image

(b) Processed image

FIGURE 8.12 Thresholding of the COVID-19-infected lung CT scan slice.

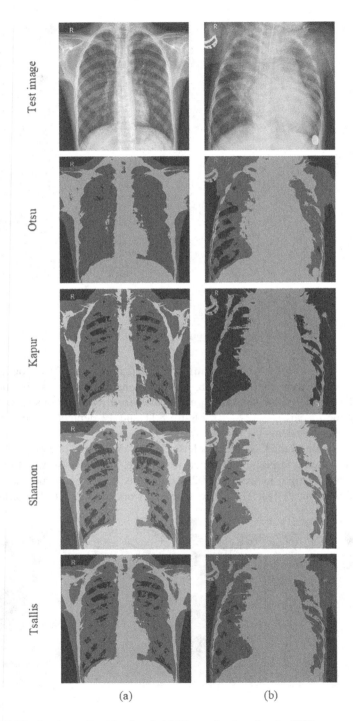

FIGURE 8.13   Implementing the threshold filter and enhancing the ROI.

COVID-19 affects the respiratory tract and causes severe pneumonia and premature detection will help to cure the patients. The common procedures followed in the clinical level detection of the COVID-19 include (i) Reverse Transcription Polymerase Chain Reaction (RT-PCR) test and (ii) Image (CT scan/Chest X-ray) assisted detection. The RT-PCR is a laboratory test initially implemented to test the collected sample from the COVID-19 patient. If the RT-PCR test result is positive, then the doctor will suggest for the image-assisted detection procedure to discover the infection and its severity. The CT scan/Chest X-ray is recorded using the radiology facility and recorded images are then directed to the doctor for further assessment. The outcome of the image-assisted detection is then considered to plan for the appropriate treatment to recover the patient from COVID-19 infection.

This section presented the enhancement of the COVID-19 infection without and with artifact removal procedure. The implemented technique works well on both the cases, and a suitable segmentation procedure can be employed to extract evaluate the COVID-19 infection for the further assessment.

8.6 SUMMARY

This chapter presented the information on the commonly used benchmark datasets and their assessment. In this chapter, medical images recorded with various modalities are examined using the CS-assisted Otsu-/Kapur-based trilevel thresholding process. The gray and RGB scaled medical images are assessed using both the Otsu's and Kapur's techniques, and from the attained result, it is confirmed that Kapur's technique is preferred for most of the medical image cases compared to Otsu's technique. The Kapur-based thresholding exactly extracts/enhances the abnormal section of the image with better visibility. In future, other entropy-based techniques, such as Tsallis, Fuzzy-Tsallis, and Shannon, are to be implemented on the datasets and the attained results.

REFERENCES

1. Rajinikanth, V., Satapathy, S.C., Fernandes, S.L., & Nachiappan, S. (2016). Entropy based segmentation of tumor from brain MR Images–A study with teaching learning based optimization. *Pattern Recognition Letters*, *94*, 87–94.
2. Dey, N., et al (2019). Social-group-optimization based tumor evaluation tool for clinical brain MRI of Flair/diffusion-weighted modality. *Biocybernetics and Biomedical Engineering*, *39*(3), 843–856.
3. Fernandes, S.L., Rajinikanth, V., & Kadry, S. (2019). A hybrid framework to evaluate breast abnormality using infrared thermal images. *IEEE Consumer Electronics Magazine*, *8*(5), 31–36.
4. Rajinikanth, V., Dey, N., Kumar, R., Panneerselvam, J., & Raja, N.S.M. (2019). Fetal head periphery extraction from ultrasound image using jaya algorithm and chan-vese segmentation. *Procedia Computer Science*. *152*, 66–73.
5. Rajinikanth, V., Satapathy, S.C., Dey, N., Fernandes, S.L., & Manic, K.S.. (2019). Skin melanoma assessment using Kapur's entropy and level set—A study with bat algorithm. *Smart Innovation, Systems and Technologies*, *104*, 193–202.

6. Manic, K.S., Priya, R.K., & Rajinikanth, V. (2016). Image multithresholding based on Kapur/Tsallis entropy and firefly algorithm. *Indian Journal of Science and Technology*, *9*(12), 89949.

7. Available online: https://wiki.cancerimagingarchive.net/display/Public/RIDER+ Breast+MRI. (Accessed on 16th August, 2020)

8. Meyer, C.R., Chenevert, T.L., Galbán, C.J., Johnson, T.D., Hamstra, D.A., Rehemtulla, A., & Ross, B.D. (2015). Data from RIDER_Breast_Mri. *The Cancer Imaging Archive*. Available online: https://doi.org/10.7937/K9/TCIA.2015.H1SXNUXL.

9. Clark, K., Vendt, B., Smith, K., Freymann, J., Kirby, J., Koppel, P., Moore, S., Phillips, S., Maffitt, D., Pringle, M., Tarbox, L., & Prior, F. (2013). The cancer imaging archive (TCIA): Maintaining and operating a public information repository. *Journal of Digital Imaging*, *26*(6), 1045–1057.

10. Pedano, N., Flanders, A.E., Scarpace, L., Mikkelsen, T., Eschbacher, J.M., Hermes, B., & Ostrom, Q. (2016). Radiology data from the Cancer Genome Atlas Low Grade Glioma [TCGA-LGG] collection. *The Cancer Imaging Archive*. Available online: http://doi.org/10.7937/K9/TCIA.2016.L4LTD3TK.

11. Chang, K. et al. (2019). Automatic assessment of glioma burden: A deep learning algorithm for fully automated volumetric and bi-dimensional measurement. *Neuro-Oncology*, *21*(11), 1412–1422. doi:10.1093/neuonc/noz106.

12. Menze, B.H., Jakab, A., & Bauer, S., et al (2015). The multimodal brain tumor image segmentation benchmark (BRATS). *IEEE Transactions on Medical Imaging*, 34(10), 1993–2024.

13. Brain Tumour Database (BraTS-MICCAI). Available online:http://hal.inria.fr/ hal-00935640. (accessed on 15 February 2020).

14. HVSMR 2016"s benchmark cardiac MRI dataset. Available online: http://segchd.csail. mit.edu/. (Accessed on 16th August, 2020)

15. Available online: http://www.onlinemedicalimages.com/index.php/en/. (Accessed on 16th August, 2020)

16. Ifan Roy Thanaraj, R., Anand, B., Allen Rahul, J., & Rajinikanth, V. (2020). Appraisal of Breast Ultrasound Image Using Shannon's Thresholding and Level-Set Segmentation. In:, Das, H, Pattnaik, P, Rautaray, S, & Li, KC (eds.), *Progress in Computing, Analytics and Networking. Advances in Intelligent Systems and Computing*, 1119, Springer–Singapore.

17. Available online: https://www.dropbox.com/s/mdx13ya26ut2msx/iChallenge-AMD-Training400.zip?dl=0. (Accessed on 16th August, 2020)

18. Available online: https://refuge.grand-challenge.org/iChallenge-AMD/. (Accessed on 16th August, 2020)

19. Fumero, F., Alayon, S., Sanchez, J.L., Sigut, J., & Gonzalez-Hernandez, M., (2011). RIM-ONE: An Open Retinal Image Database for Optic Nerve Evaluation. In: *24th International Symposium on Computer-Based Medical Systems*, 1–6, IEEE: Bristol, UK.

20. Available online: https://www.fc.up.pt/addi/ph2%20database.html. (Accessed on 16th August, 2020)

21. Glaister, J., Wong, A., & Clausi, D.A. (2014). Segmentation of skin lesions from digital images using joint statistical texture distinctiveness. *IEEE Transaction on Biomedical Engineering*, *61*(4), 1220–1230.

22. Prinyakupt, J., & Pluempitiwiriyawej, C. (2015) Segmentation of white blood cells and comparison of cell morphology by linear and naïve Bayes classifiers. *BioMedical Engineering OnLine*, *14*, 63.

23. Bernal, J., Tajkbaksh, N., Sánchez, F.J., Matuszewski, B., Chen, H., Yu, L., Angermann, Q., Romain, O., Rustad, B., Balasingham, I., Pogorelov, K., Choi, S., Debard, Q., Maier-Hein, L., Speidel, S., Stoyanov, D., Brandao, P., Cordova, H., Sánchez-Montes, C.,

Gurudu, S.R., Fernández-Esparrach, G., Dray, X., Liang, J., & Histace, A. 2017. Comparative validation of polyp detection methods in video colonoscopy: Results from the MICCAI 2015 endoscopic vision challenge, *IEEE Transactions on Medical Imaging*, *36*(6), 1231–1249.

24. Available online: https://radiopaedia.org/cases/covid-19-pneumonia-27. (Accessed on 16th August, 2020)
25. Available online: https://www.nejm.org/coronavirus. (Accessed on 16th August, 2020)
26. Available online: https://www.who.int/emergencies/diseases/novel-coronavirus-2019://www.who.int/news-room/fact-sheets/detail/pneumonia. (Accessed on 16th August, 2020)

Index